The Origin of Life

The Origin of Life

Toward a Theoretical Biology

C.H. Waddington, editor

Volume 1

Routledge
Taylor & Francis Group
LONDON AND NEW YORK

First published 2009 by Transaction Publishers

2 Park Square, Milton Park, Abingdon, Oxfordshire OX14 4RN
711 Third Avenue, New York, NY 10017

Routledge is an imprint of the Taylor & Francis Group, an informa business

First issued in hardback 2017

Library of Congress Catalog Number: 2008027964

Library of Congress Cataloging-in-Publication Data

[Towards a theoretical biology]
The origin of life / [edited by] C.H. Waddington.
p. cm.
Originally published: Towards a theoretical biology / edited by C. H. Waddington. Aldine Pub. Co., 1968-
Includes bibliographical references and index.
ISBN 978-0-202-36302-8 (alk. paper)
1. Biology--Philosophy--Congresses. 2. Biology--Congresses. I. Waddington, C. H. (Conrad Hal), 1905-1975.

QH301.T68 2008
570.1--dc22

2008027964

ISBN 13: 978-0-202-36302-8 (pbk)
ISBN 13: 978-1-138-53725-5 (hbk)

Contents

Preface

Theoretical Physics is a well recognized discipline, and there are Departments and Professorships devoted to the subject in many Universities. Moreover it is widely accepted that our theories of the nature of the physical universe have profound consequences for problems of general philosophy. In strong contrast to this situation, Theoretical Biology can hardly be said to exist as yet as an academic discipline. There is even little agreement as to what topics it should deal with or in what manner it should proceed; and it is seldom indeed that philosophers feel themselves called upon to notice the relevance of such biological topics as evolution or perception to their traditional problems.

The International Union of Biological Sciences has felt that it is its duty, as the central focus of international organizations of all the branches of biology, to explore the possibility that the time is ripe to formulate some skeleton of concepts and methods around which Theoretical Biology can grow. It was clear that the task would be by no means easy; and it was therefore arranged that a series of three Symposia should be held at yearly intervals. The intention was that the discussions would be concerned, not with the theory of particular biological processes, such as membrane permeability, genetics, neural activity, and so on, but rather with an attempt to discover and formulate general concepts and logical relations characteristic of living as contrasted with inorganic systems; and further, with a consideration of any implications these might have for general philosophy. I was asked to invite suitable speakers and to organize the meetings.

The first Symposium was held from 28 August to 3 September 1966, at the Villa Serbelloni, Bellagio, Lake Como, at the kind invitation of the Rockefeller Foundation. As a preparatory document, to raise some of the problems, I circulated an intentionally provocative version of a series of Ballard Matthews Memorial Lectures given earlier in the year at the University College of North Wales, Bangor. Some comments by René Thom on these lectures were also circulated before the meeting, as was an article by Ernst Mayr.

The discussions at the Villa Serbelloni, which were very free, intense, and stimulating, were not recorded. They tended to focus on the problems of biological theory rather than on the more philosophical issues. Although some glimmerings of a firm framework for theoretical biology seemed to be emerging, it was clear that more discussion and interchange between people of very different viewpoints would be required before anything like an outline of an academic discipline could be drawn. The bulk of this book therefore consists of separate essays written after, and in the light of, the Symposium. They are not yet connected with one another into any kind of a unity. It was, after all, the recognition that no such unity exists, and that bringing it into being would be a long and difficult task, that led the IUBS to plan for three successive Symposia. The intention is that at the second Symposium (1967) some further progress towards a synthesis may be achieved. It is towards this goal that these results of the first Symposium are offered as PROLEGOMENA.

The Executive Committee of the International Union of Biological Sciences, which proposed the Symposium, and the scientists who took part in it, both owe a deep debt of gratitude to the Rockefeller Foundation for making the Villa Serbelloni available to them. The beauties of the Villa and its surroundings, and the gracious hospitality of Dr and Mrs Marshall, provided exactly the right context for informal but penetrating discussions. The Royal Society of London and the Office of Naval Research of the United States also contributed generously to the expenses.

I should like to express my personal thanks to the Principal and Senate of the University College of North Wales for permission to publish the Ballard Matthews Lectures in this form. Thanks are also due to the editors of *Science* and *Nature*, who have given permission for the reprinting of the articles by Mayr (with comment by Waddington), and by Michie and Longuet Higgins.

C. H. WADDINGTON
University of Edinburgh

The Basic Ideas of Biology

C. H. Waddington
University of Edinburgh

I shall try in these lectures* to step back a little distance from all the detailed problems presented by living systems, even though they are so fascinating that they can keep a working biologist enthralled throughout a long active life. What I want to do now is to look at biology as a whole and try to discern what are the basic ideas characteristic of this type of science. It is probably best to begin boldly by asking the brash question, 'What is life ?'. In the fairly recent past most authors have echoed the sentiments of Sherrington about this : 'To ask for a definition of life is to ask a something on which proverbially no satisfactory agreement obtains.' However, this did not entirely prevent them giving their opinions about the matter. And in recent years the question has become very acute again. When we obtain actual specimens of the surface constituents of the moon or Mars or even fairly detailed televised information about them, what sort of evidence would we accept as establishing the existence of anything worthy to be called 'life' ?

During at least the first third of this century the most authoritative biologists saw the fundamentals of life in the characteristic ways in which living organisms operate. A living thing, they pointed out, takes in a number of relatively simple physico-chemical entities or molecules from its surroundings (incident light energy, water, inorganic salts, relatively simple organic food materials, etc.) and builds these up into its own structure, in the process carrying out a synthesis of molecular structures much more complex than those it had absorbed. 'The constant synthesis, then, of specific material from simple compounds of a non-specific character is the chief feature by which living matter differs from non-living matter' was the way it was expressed by Jacques Loeb in 1916. The same point of view was put by another leading biologist of that day who represented quite the other end of the spectrum from Loeb in relation to the controversy about vitalism and mechanism, namely J. S. Haldane : 'The active maintenance of normal and specific structure is what we call life, and the perception of it is the perception of life. The existence of life as such is thus the axiom upon which scientific biology depends.'

This view of the basic nature of biology was the inspiration of the conscious

* The Ballard Mathews lectures, University of North Wales, Bangor.

movement for the development of a 'modern biology' which took place in Great Britain at the end of the 1920's and early 1930s. This took as its aim the study of 'the nature of living matter', which was the title of a book by one of its leaders, Lancelot Hogben, and it focused on Loeb's idea that there is a special type of substance, living matter. Some authors simply identified living matter with the cell. Sherrington, a few lines after the quotation given above, goes on: 'There is a little aggregate of atoms and molecules such as the world we call lifeless nowhere contains. It is confined to living things; many of these it enters into as a unit, and builds up. It is a unit with individuality. . . .' To these units Robert Hooke of the Royal Society, an early observer with the microscope, gave the name 'cells'. The most influential leader of the modernists, James Gray, states that 'It seems logical to accept the existence of matter in two states, the animate and the inanimate, as an initial assumption.' However, he did not suppose that the cell is the unit of matter in the animate state. His reasons for this were partly empirical, the fact that one can find rather exceptional organisms in which the cellular organization is not developed (for instance, multinucleates, algae or slime moulds) but, more importantly, because the cell is such a comparatively large entity containing so many molecules that it must be subject to statistical regularities and thus under the control of the second Law of Thermodynamics. One of the most profound characteristics of living systems, regarded as working machines, is that they lead to a local increase of order by taking in simple molecules and building them up into complex compounds which are arranged in an orderly fashion. At first sight, this might seem to imply that the system acts against the dictates of the second Law (cf. Kornacker's article, p. 94). Gray therefore supposed that 'the real unit of life must be of a protoplasmic nature irrespective of whether it is sub-divided to form a mechanically stable system or not: in other words, cellular structure is not in itself of primary significance. . . . It is by no means impossible, however, that the essential units of living matter are each composed of a very large number of molecules and under such circumstances the number of units present may be reduced to a figure which will materially affect the validity of statistical laws.' As these quotations make clear, the refusal to accept the cell as the basic unit of life left intact the notion that life is to be defined as a particular state of matter.

This conception underlay the growth of modern biology conceived of as the experimental study of living organisms as causally operating systems. Within a few decades there was a great increase in our understanding of such processes as muscular contraction, glandular secretion, nervous conduction, the metabolic transformation of molecules one to another, and the whole functional machinery by which a living thing works. It gradually became clear that in nearly all these processes the fundamental role is played by the proteins, which are enormously complex molecules that are found only in living (or recently dead) biological entities. Perhaps the greatest triumph of this type of biology has been to elucidate their structure. It was first shown that they are built up by attaching smaller molecules of amino acids in sequence one to another to form a long string-like structure. Each such string of amino acids is folded together into a complex tangle, and the effective operations of the protein depend essentially on the precise shape that this tangle takes. It is only in the last few years, thanks to the work of X-ray crystallographers such as Astbury, Pauling, Bernal, Perutz, and Kendrew, that we are at last beginning to get precise information about the shapes of these tangles, the forces that hold them together, and the degree to which they can be modified. If life were really no more than a particular state of matter, these studies would be coming very near to giving us an understanding of the basic factors on which the 'nature of living matter' depends.

However, it was not long before the adequacy of the 'living matter' definition of life was seriously challenged. It was pointed out, chiefly by geneticists, that living things do not merely synthesize specific structures out of simpler molecules; it is an equally important fact that they reproduce themselves, and indeed it might be claimed that the most important fact about them is that they take part in the long-term processes of evolution. The capacity to reproduce, that is to give rise to a new unit essentially similar to the old one, demands not merely specific synthesis but the ability to pass this specificity from the initial unit to a new unit which is its offspring. For evolution to be possible something still further is required. It is necessary that changes occur from time to time in the specificity of an organism and that, when such changes occur, they are passed on to the offspring. This will amount to the occurrence of hereditary variation, and Darwin's argument for the inevitableness of natural selection will then be sufficient to ensure that evolution will take place. H. J. Muller, the most farsighted of the early geneticists, argued as far back as 1916 that this provided the only basis for a satisfactory definition of life. A system is living if it carries specificity and can transmit this specificity to offspring and if, in addition, the specificity can change and the changed specificities are also transmitted. Nowadays, using modern jargon, we should rephrase this slightly, using the word information instead of specificity: a system is living if it encodes hereditarily transmittable information, if this information sometimes suffers alterations, and if the altered information is then transmitted.

The geneticist's definition of life is obviously much more general than the physiologist's definition given above. It does not connect life essentially with the particular types of living matter which we can inspect on the earth's surface. In point of fact on the earth's surface hereditary information is, so far as we know, always encoded in a class of compounds known as the nucleic acids and, in our experience, this is always expressed by the formation of corresponding proteins which carry out the functions which the physiological biologists have investigated and described. But the genetical definition does not require that this should be so. We might have genetic information encoded in other compounds and worked out in a different way. This opens a much wider range of possibilities when we contemplate the possibility that there exists something worthy to be spoken of as 'life' on other planets or stars in which the physico-chemical conditions are totally different from those on the surface of the earth. Moreover, it makes it possible to discuss, in a meaningful way, the problem of the origin of life from inorganic materials. For these reasons, it has, in recent years, tended to be more and more widely adopted, and the old idea that there is a characteristic type of 'living matter' has passed more and more into the background.

In my opinion, this reaction, although basically justified, is in danger of going too far. I do not believe that if we discovered systems that were nothing more than mechanisms for the hereditary transmission of mutable information we should, in fact, consider them to be living. Such systems would show none of the properties included in the older physiological outlook, which regarded the characteristic of life as specific synthesis and an apparent local reversal of the second Law of Thermodynamics. One can get some idea of their character by considering the well-known game in which you are given a word, say Bit, and are required to change this into something else, say Man, by altering one letter at a time in such a way that the new combination always spells an accepted English word; and we could, if we liked, expand the rules so that letters could be duplicated or pieces of the word reversed in order or transposed from one place to another, etc. We have perhaps a system rather like this in the growth of crystals. In a normal perfectly formed crystal new molecules are laid down in regular order on the face of a pre-existing arrangement of molecules, but if a mistake happens and one new molecule gets into a slightly wrong position this frequently produces an irregularity on the surface which affects the probability that a new molecule will settle down at that place. Thus the mistake is, in some sense, transmitted in that it affects the next generation of molecules. If the irregularity is such as to increase the chance of a new molecule being deposited there, one could regard this as a selectively favourable effect. We can thus have processes essentially akin to the natural selection of mutable hereditarily transmitted information, but we do not regard such systems as alive, and I think the reason is essentially because they are not interesting enough ! They fulfil Muller's criterion but they do absolutely nothing else. To be worthy of being called alive they must, I think, exhibit some sort of 'physiological activity'. This need not of course be of the kind exhibited by living things on this earth. They might produce 'living matter' of a totally different chemical composition and mode of operation from that which we find around us, but I should argue they need to do something more than mere information-transmission.

Muller defined life entirely in terms of what geneticists call the genotype. The argument I am putting forward amounts to saying that life involves not only the genotype but also the production of something of the kind geneticists speak of as the phenotype, that is to say, something which was developed out of the genotype and which interacts with the surrounding non-living environment. It was the phenotypes produced on earth that the physiologists were considering when they defined life in terms of a state of matter. Although one cannot accept this geocentric limitation, there is, I think, a good case for insisting that a living system must have some sort of phenotype or other.

THE THEORY OF PHENOTYPES

Whether or not we accept the argument that life necessarily involves not only genotypes but also phenotypes, there is no doubt that all the living systems we know on this earth do have a phenotype as well as a genotype – even viruses. The theory of phenotypes is therefore an essential part of the general theory of biology. It will be discussed under a number of separate headings.

▸ *Information theory.* Around 1948 Shannon and Weaver developed an elegant mathematical theory for handling certain problems in which the point at issue was the amount of variety or specificity contained within a system. There are very many biological problems where we might be tempted to use this same form of words – amount of variety or specificity. For instance, we have seen that the physiological definition of life was phrased in terms of 'specific synthesis'. Again, we speak of genetic variation, or say that the genotype specifies the nature of the phenotype. There has therefore been a great temptation to use the Shannon-Weaver theory in connection with biological problems. However, their 'information theory' was developed in connection with a particular type of process and has limitations which make it extremely difficult, if not impossible, to use it in many of the biological contexts to which people have been tempted to apply it.

The theory was developed (in the Bell Telephone laboratories) in connection with the problem of transmitting a message from a source A through a channel B to a receiver C, and the basic purpose for which it was evolved was the question of how the characteristics of the channel B influenced the amount of information that can be transmitted in a given time. One of its basic results is that, in a *closed system* into which nothing comes from outside, you can never get more information into C than was originally contained in A, although you can, of course, change the form of information, say, from the dots and dashes of the Morse code into letters written with a typewriter. Now there are several biological situations which present a close parallel to the systems investigated by Shannon and Weaver. The most obvious example is the passage of electrical impulses through networks of nerves, where one has an almost precise parallel to the transmission of electrical impulses along telephone wires. Here their theory and its elaborations have proved extremely valuable. A slightly more far-fetched analogy can be found in the transmission of the hereditary information contained in the chromosomes of one organism to the chromosomes of its offspring. But here already we find that biology has developed mechanisms more flexible than those used by telephone engineers. There is a system by which the transmitted information can be changed – a gene may mutate so that what the offspring receives is not exactly the same as was contained in its parent. Again, there are mechanisms of chromosomal deficiencies, duplications, translocations, formation of isochromosomes, etc., by which the amount of information can be either increased or decreased. However, the information theory language is useful insofar as it allows these situations to be clearly expressed.

It is when we come to consider the relations between the genotype and the phenotype that the limitations of the theory become of overriding importance and rapidly render it not merely useless but a dangerous snare. In the very first steps of the transition from genotype to phenotype it can still be applied. The genes, composed of D N A, begin to make their effects felt by serving as the patterns (templates) on which there are synthesized the messenger R N As. In these there is, for every sugar-nucleotide unit in the D N A, a precisely corresponding sugar-nucleotide unit in the R N A. The change is no more than, as it were, changing from a Roman face type to an Italic type, using the same alphabet. It is usually spoken of in biology as a 'transcription'. Even in the next step there is nothing very much to be alarmed at. The sequence of sugar-nucleotide units in messenger R N A provides the pattern for the laying down of a corresponding sequence of amino acids into a polypeptide chain that forms a protein. This involves a more drastic alteration, perhaps comparable to changing from a normal alphabet to one in Morse code. It is usually spoken of in biology as a 'translation'. It is meaningful to discuss the whole of this sequence in terms of information theory and to ask whether any information is lost in the processes of transcription or translation, and to pose questions about the 'code' according to which the RNA nucleotides are translated into the protein amino acids.

It is in the next stages of the formation of the phenotype that information theory becomes unable to deal with the situation. It is obvious that the phenotype of an organism does not consist simply in a collection of all the proteins corresponding to all the genes in the genotype and nothing else. It is instead made up of a highly heterogeneous assemblage of parts, in each of which there are some, but not all, of the proteins for which the genes could serve as patterns, and in each of which there are also many other substances and structures over and above the primary proteins corresponding to particular genes. Each organ – liver, kidney, brain, etc. – is in the first place like a Bridge hand, containing only a certain sample out of the whole pack of cards, and further, contains a whole lot of things – lipids, carbohydrates, pigments, and what have you – that are secondary accretions which one cannot consider as having been in the original hand when it was first dealt. Information theorists have struggled in vain with this problem. It seems quite obvious to common sense that a rabbit running round in a field contains a much greater 'amount of variety' than a newly fertilized rabbit's egg. How are we to deal with the situation in terms of an 'information theory' whose basic tenet is that information cannot be gained? There have been several attempts:

(a) Raven has argued that we have underestimated the amount of information contained in the egg because we have considered only the information contained in the chromosomes, i.e. the genes. He argues that there is also information contained in the cortex of the egg. Taking the area of this cortex and an estimate of the size of a biological molecule, he comes up with another quantity of information which can be multiplied into that in the chromosomes, and he tries to show that this is not too far away from the amount of information contained in the adult organism, as estimated by taking its size in terms of the size of the molecules of which it is composed. But the whole set of estimates here is so fantastically imprecise – Raven considered the cortex of a small egg but could have got himself another factor of 10^4 or 10^5 if he had taken an egg of the size of a hen's ovum – that the whole procedure becomes totally unconvincing.

(b) One can also point out that the system we are dealing with is not a closed one. An organism, as it develops from a fertilized zygote to an adult phenotype, nourishes itself by taking in food from its surroundings. Now the molecules of food – proteins, amino acids, fats, carbohydrates, etc. – all contain some chemical information over and above the bare atoms of which it is composed. Could the adult rabbit have got its extra information, additional to that in the zygote, from the grass it ate? I frankly don't know the answer, but it sounds to me like suggesting that you could add to the information in a paragraph of prose by using a more curly serifed typeface instead of simple sanserif designed on the basis of nothing but straight lines and segments of circles.

(c) But the basic point is surely that in the transition from the zygote to the adult the 'information' is not merely being transcribed and translated but is operating as instructions – if you want to put it in fancy jargon, as 'algorithms'. The DNA makes RNA and the RNA then makes a protein and the protein then does some thing to its surroundings, which results in the production of more varieties of molecule than there were before. There is nothing very mysterious in this, unless you try to see it in terms of messages going down telephone wires. Suppose you put together in a reaction vessel two chemical species A and B, where A is a molecule having NH_2 groups at a few sites – say, 1, 3, 9 – and B is a molecule with the ability to substitute an OH group for an NH_2 group. After a time there will be a much greater variety of molecules in the reaction vessel than there were to start with. The amount of 'information' will have increased, but there is no reason why we should feel that we are witnessing a phenomenon which contravenes the fundamental laws of the physical universe. All we need to realize is that we are dealing with a situation outside the very narrow limits to which information theory applies.

This is, so far as I can see, all that is meant by Elsasser's argument that a comparison of the information content of a fertilized egg and adult phenotype makes it necessary to postulate the existence of 'biotonic laws' – what this boils down to is that information theory cannot usefully be used in considering the relation between genotypes and phenotypes. As I put it some years ago, a genotype is like a set of axioms, for instance Euclid's, and a phenotype like a three-volume treatise on Euclidean geometry, which proves Pythagoras's theorem in the liver and in the kidney that a tangent to a circle is perpendicular to the radius at the point of contact, in the lungs that the three angles of a triangle add up to 180°, and so on. The inapplicability of information theory to such situations may be demonstrated by pointing out, firstly, that a theorem is proved by showing that it is tautological expansion of the axioms and therefore contains nothing that was not in the axioms to begin with, but that, secondly, giving such a demonstration is by no means without significance, since if you could prove Pascal's theorem that $a^n + b^n = c^n$ is impossible for any value of n greater than 2 you would be assured of undying fame among the tribe of mathematicians.

In sum, information theory in the strict sense is not a useful tool in considering the relation between the genotype and the phenotype, although the word 'information' is, if used fairly loosely, quite a useful expression to employ in place of such phrases as 'amount of variety' or 'specificity'.

▸ *The phenotype as an epigenetic phenomenon.* At the beginning of all textbooks of genetics, a very necessary distinction is drawn between the genetic potentialities which an organism inherits from its parents and the manner in which these potentialities become actually realized. The former is the genotype, the latter the phenotype. It is common, even in the best and most recent textbooks, to describe the phenotype in an extremely incomplete and over-simple manner. For instance, Srb, Owen, and Edgar are content with saying: '*Phenotype*, a word that refers to the appearance of an individual'. Even such a cautious and farsighted author as Stern, after pointing out that 'A character or trait may be defined as any observable feature of the developing or fully developed individual', goes on to say 'For the genetic constitution, the term *genotype* has been coined; for the external appearance, the term *phenotype*.' Mayr comes nearer to getting the matter into proper perspective when he writes: 'Our ideas on the relation between gene and character have been thoroughly revised and the phenotype is more and more considered, not as a mosaic of individual gene-controlled characters, but as the joint product of a complex interacting system, the total epigenotype (Waddington, 1957).' Even Mayr, however, does not clearly express the most fundamental and basic characteristic of phenotypes, namely that they change in time. The phenotype is the name given to the results of the activities of genes. In a very simple organism, these activities may be carried out in a relatively short period, resulting in the formation, for instance, of a certain number of proteins. In more complex organisms, these proteins themselves interact with one another and with other substances so that it is a long and complex sequence of processes that the genes set in motion. But in either case a time duration, whether short or long, is an essential component of a phenotype.

One can use the expression 'a phenotypic character' (or character for short) to refer to any particular aspect of a phenotype which is being singled out for attention. Some years ago (e.g. 1947) I introduced the word 'epigenetics', derived from the Aristotelian word 'epigenesis', which had more or less passed into disuse, as a suitable name for the branch of biology which studies the causal interactions between genes and their products which bring the phenotype into being. The word is now quite often used in that sense; but unfortunately it seems to be an attractive expression, and some other authors have seized on it to refer to quite different conceptions. For instance, Elsasser wants to use the adjective 'epigenetic' to indicate that in a certain process information is not being conserved according to the orthodox Shannon-Weaver theory. In my opinion it would lead to greater clarity if the word is kept for the causal study of development, the meaning originally suggested for it.

If one wishes to formulate the phenotype in mathematical terms it is clear, then, that it is a function which involves the time variable. Moreover, the function must involve something more than merely the three dimensions of space, since we are interested in something more than the bare geometry of the organism. We shall need in fact one variable for each constituent (chemical or geometrical) of the system which is relevant to the questions being considered. We shall therefore usually be faced with a function containing very many variables. In order to represent this geometrically we should have to resort to a multi-dimensional phase space to accommodate all the variables of constitution, and within this space the phenotype will be represented by some kind of figure, which will begin at the point representing the constitution of the egg and will then be extended along the time dimension. Theoretically this figure might take the form of a bounded continuous sheet, for instance that of a triangle. If this were the case, one should find that, at some time after fertilization, the phenotype exhibited continuous variation in composition as one passed from one position within it to another. It is an empirical observation that this is normally not the case. In the organisms that we come across, we usually find a number of discrete and distinct organs – a liver clearly marked off both spatially and in composition from the kidney, and both of these from the heart, and so on. This means that the figure representing the phenotype has to branch out into a number of separate sub-configurations, each of which extends separately forward along the time dimension. Little generality is lost if one represents each of these sub-configurations by a line. One can therefore represent the phenotype as a branching system of time-extended trajectories in the phase space.

Although these points may seem obvious, and were stated as long ago as 1940, it is doubtful how far they yet form a part of the working system of thought of most biologists. For instance, one still often sees the remark that cell differentiation during development is explained by the switching of cells into 'alternative steady states', the idea being often attributed to Delbrück. But if one considers any actual case, such as the switching of part of the ectoderm of an amphibian gastrula to become neural tissue or to become epidermis, it is quite clear that it does not enter a steady state but, on the contrary, sets off on some particular pathway of change which, for instance, in one direction will lead it to form part of the neural plate, then the neural tube, then some region of the spinal column, or brain forming part of the grey matter or white matter or perhaps growing out as a nerve trunk. It is only after it has passed through a large number of transient stages that it gets into a situation which is relatively stable for a considerable period of time, namely the adulthood of the individual.

▸ *The elementary processes of epigenetics*. Epigenetics has two main aspects: changes in cellular composition (cellular differentiation, or histogenesis) and changes in geometrical form (morphogenesis). How far can we specify what are the elementary processes to which the more complex examples of these phenomena should be reduced?

1. About morphogenesis we can, I believe, say very little. Right at the bottom of the 'ladder of causes' there must be such processes as:

(a) the determination of tertiary structures of proteins by the nature of their primary amino acid sequence,

(b) the association, often by still weaker types of bonding, of macromolecules (possibly sometimes with ordinary molecules as mortar between the joints) such as we see it in the reformation of striated collagen or myosin fibrils from solution, the formation of myelin figures or other sheet-like structures, in systems involving immiscible phases, etc. But there is an enormous gap between such phenomena and, for example, the appearance of just five digits of particular lengths and shapes in a developing limb. I do not think we can indicate any one type of elementary process (or even a small number of them) which provide a general pattern for the intermediate steps. It seems rather likely indeed that these steps are carried out by a large number of different types of mechanism and that there is no one unitary theory of the way morphogenesis is brought about.

2. We can probably go somewhat further in relation to cellular differentiation. It is in fact conventional to say that the basic elementary process is the derepression (or possibly switching on) of a structural gene by means of a cytoplasmic gene-recognizing ('genotropic') substance which has been produced by some other genetic locus. The notion of the derepression of single genes has been derived primarily from work on bacteria, and the question arises whether or not there is another more complicated level of elementary process which is basic for the cells of higher organisms, in which the DNA of the chromosomes is normally combined

with protein. As far as I know, there is no evidence that you can ever switch on single genes in higher organism cells, except in the terminal phases of development when a lot of other correlated genes have already been switched on (e.g. you can switch on or off the haemoglobin gene in cells of the erythropoetic series but not in kidney cells or nerve cells).

I have argued that the elementary differentiation process in a higher organism cell (a) involves complexes or 'batteries' of genes rather than single genes; (b) takes place in three phases: acquisition of competence, in which several different batteries of genes become ready to enter the next phase; determination, in which one of these batteries is singled out to become dominant in the future history of the cell; activation, in which the proteins corresponding to the structural genes in this battery actually begin to be produced. Since cells can divide while remaining in phase 2 (determined but not activated) it seems almost certain that determination must be a process operating at the gene level. It is not clear, however, whether it leads to the production of corresponding messenger R N As, or whether this is always the first phase of activation; even if the latter is the case it must be admitted that complete activation may also involve control of protein synthesis at the ribosome level.

▸ *The canalized or buffered character of the epigenetic trajectories.* It is an empirical observation – but one of profound importance for much of biological theory – that epigenetic trajectories normally show some resistance to being changed. The evidence is of two main kinds:

(a) Developmental: It is extremely common to find that a developing system is, at least for some periods of time, capable of 'regulation', in the sense that it is capable of 'compensating for' disturbing influences and returning to normality at later stages in the developmental process.

(b) Genetic: It is again very usual to find that slight changes in a genotype may produce no deviation in the developing phenotype (e.g. in the phenomena of dominance, epistasis, etc.).

Phenomena involving the holding constant of some parameters of a physiological situation (e.g. the oxygen tension or pH of the blood) have been well known for a long time. The situation is usually referred to as one of 'homeostasis'. We are here dealing with a similar concept, but of a rather more general nature, in that the thing that is being held constant is not a single parameter but is a time-extended course of change, that is to say, a trajectory. The situation can therefore be referred to as one of homeorhesis, i.e. stabilized flow rather than stabilized state.

(I should like to see some mathematician express this contrast in more precise terms; for instance, homeostasis does not in general involve the system returning to a single point in phase space. In order to keep constant the oxygen tension of the blood you may increase the rate of heart-beat. I suspect that the situation is something like this: in homeostasis the phase space contains what I think Thom would call an 'attractor line', in the equation for which one variable had a constant value; in homeorhesis the attractor line is parallel to the time axis and none of the variables are necessarily constant.)

The name 'chreod' has been suggested to refer to a canalized trajectory which acts as an attractor for nearby trajectories. It is an interesting question to discuss how far the existence of chreods is necessary and how far it is merely an empirical result of the operation of natural selection. There are several points which I do not quite know how to put together:

(i) In many of the chreods we meet in actual animals, both the value of the parameters and the degree of stabilization is under the control of natural selection. For instance, in selection experiments involving body size in Drosophila, Forbes Robertson found good evidence for the buffering of final body size to particular values, but the actual magnitude of these values was different in different selection lines. Again, it is a common observation that the introduction into a relatively normal genotype of some strongly acting mutant (which does not persist in natural populations so that its effects can be subject to natural selection) often produces a phenotype which is not only highly abnormal but also highly variable, i.e. it destabilizes the chreod. Finally, artificial selection can certainly build up new chreods, as in the experiments on genetic assimilation.

(ii) However, this does not seem to me to rule out the possibility that the existence of some sort of chreod or another is a necessity. If one mixes together a large number of active agents (say chemical substances) which can act upon each other, the mixture will change through some defined course in time. The question is, will this course show any properties of buffering or canalization? I should have guessed intuitively that, provided a large number of the components can interact not only pairwise but each with a number of other components, some degree of buffering would be bound to emerge.

(iii) When any of the interactions between the components are strongly non-linear – crudely speaking, involve threshold phenomena – then some sort of chreodic behaviour seems to be inevitable (see further discussion of this in the section on archetypes in evolution).

Another set of questions concerns the type of mechanism by which the buffering is produced. It is becoming conventional to refer in this connection to 'feedback'.

Strictly speaking, as I understand it, feedback refers to a situation in which, if process A deviates from normal, the system reacts in such a way as to bring process A back to normality. In biology something of this kind does occur in such processes as end-product repression and end-product inhibition, in which the final product of an enzymatic pathway acts so as to control various earlier steps in the sequence by which it is synthesized. However, there is another and perhaps more usual type of buffering action in which, when process A deviates from normality, nothing happens to bring it back again, but the system as a whole simply absorbs its effects. For instance, Kacser has shown that if a mutation drastically reduces the efficiency of one enzyme in a sequence of synthetic steps, the consequence may merely be an increase in the concentration of its immediate precursor, leading to the same overall flow rate through that synthetic step and complete absence of any effect on the later steps. Much more complex compensatory reactions could, of course, be envisaged in networks, rather than simple sequences, of chemical reactions. This is absorptive buffering rather than feedback in the strict sense.

There is, however, no doubt that feedback dependent on direct coupling between processes does often occur. In particular, the regulator-operon system of biochemical linkage between genes is likely to give rise to many 'strong' interactions between cellular processes. The general theoretical consequences of these have been investigated by Goodwin. As he will be at the Symposium I shall not attempt to summarize, except to point out that his argument that oscillatory phenomena are to be considered as the normal pattern of cellular biochemistry seems to me to open a great many new avenues of thought in fundamental general biology.

The importance of understanding the nature of the mechanisms on which chreodic behaviour depends arises from the fact that a chreod, as we have defined it, is simply the most general description of the kind of biological process which has been referred to as 'goal directed'. The nature of such processes has always been recognized as one of the major problems of theoretical biology. The words to be used for describing them and discussing them are still matters for debate. The earlier expressions 'teleological' and 'finalistic' are usually thought to carry an implication that the end state of the chreod has been fixed by some external agency and that the end state is in some way operative in steering the trajectory towards itself. To avoid such implications I have spoken of such phenomena as 'quasi-finalistic', and the word 'teleonomic' (introduced I believe by Pittendrigh in *Behaviour and Evolution*, 1958) has been used as a substitute for teleological.

On the whole, however, I believe it is preferable to use words (such as chreod) which do not lay such stress on the final state but draw attention to the whole time trajectory.

The two main discussions of the topic in recent years by professional philosophers of science are, I think, those by Nagel and Sommerhof. These are both set out in fairly technical language, but when one penetrates this one finds, first, that both of them have discussed situations of a quasi-finalistic kind, i.e. concentrating on the attainment of an end state rather than on following a trajectory; and secondly, so far as I can see, the crucial problem is merely smuggled-in as one of the assumptions underlying the logical algebra without being actually explained in any way. For instance, Nagel assumes the existence of two sets of 'state variables' B and C which are so related that variations in one are balanced out by corresponding variations in the other. Sommerhof indulges in essentially the same procedure with his two variables R and E. To my mind these amount to no more than statements of the problem, not explanations of them; statements moreover in a form which combines the disadvantages of being so abstract as to give no hint of a direction in which an explanation might be sought, and of being of such narrow scope that it omits the fact that we are dealing with trajectories not merely with end states. I think therefore that the plain language discussion given above penetrates further than any of the existing 'metamathematical' discussions, though it will I hope soon be superseded by some more sophisticated treatment (possibly derived from topology?).

Great hopes are often held out for the application to biological problems of new mathematical approaches grouped under the general heading of 'Systems Theory'. I hope some of the mathematicians at the Symposium can clothe these airy promises with some real flesh and blood. Looking at articles such as, for instance, that by Quastler, the sceptical biologist is likely to ask, firstly, what is the real gain in either flexibility or precision in substituting the terminology input – black box – output for the more old-fashioned stimulus – cell – response? And secondly, what actual theories in any way applicable to biology have emerged from any of the four major types of systems theory he refers to: cybernetics, games theory, decision theory, and communications theory? In too many of the articles written for biologists on such subjects (including Quastler's) the exposition stops after problems have been formulated in a new fashion and before any theorems have been proved (apart perhaps from some points about the minimax strategy in games theory which Lewontin has applied to some problems of evolution theory).

▶ *The essentially oscillatory nature of phenotypic states.* It is well known that systems involving negative feedback tend to go into oscillation; in fact one has to find reasons why they do not oscillate rather than provide reasons why they should. Goodwin has recently examined the statistical mechanical properties of biological systems on the assumption that the cell contains many components coupled together by feedback relations. His studies issue in the conclusion that such systems are inherently oscillatory. If this is so, it represents, in my opinion, a major concept in theoretical biology. Practically nobody in the past has, so far as I know, conceived of a cell as being *essentially* an oscillator, and the idea opens a whole new range of possibilities concerning the interactions between cells. At this time I should like to make only two remarks:

(i) As I understand it, the algebraic analysis is only explicitly soluble if one makes the simplifying assumption that there are only binary interactions between constituents. This seems to amount to the assumption of what I called above 'feedback in the strict sense'. Does the conclusion about oscillatory behaviour still hold if the canalization of a chreod depends on absorbtive buffering which involves interaction between very many components?

(ii) It is not obvious that the new system of ideas, important though they may be in theoretical biology, have any major implications for general philosophical questions.

▶ *The nature of biologically transmissible information.* The phenotype can be regarded as, in the main, an exposition into a number of different propositions of the information transmitted in the DNA of the chromosomes. We know of some, probably minor, variants of this type of information:

(i) Pro-virus-like bodies also consisting of DNA, which may exist either attached to the chromosome or free from it.

(ii) RNA viruses in which RNA shows the same capacity for self-replication, i.e. for transmitting information, as DNA does. Must we conclude then that all biological transmission of information is carried out by nucleic acids?

The first question is, have we good evidence for the biological transmission of information by systems lying outside the chromosome? The answer seems to be that we have. Some of this is concerned with the very highly specialized cortex of the ciliates (e.g. Sonneborn on Paramecium) or other specialized cell organelles such as chloroplasts. In both these instances there is a good deal of evidence that nucleic acids are present in the structures, and it *might be* that it is nucleic acid which transmits the information. There is other evidence of non-chromosomal heredity, particularly that of Sager, in which the location of the transmitting system is at present unknown. Again it *might be* that the transmitting agent is a nucleic acid. Indeed, the hereditary system shares some of the detailed properties of chromosomal heredity (e.g. intra-allelic recombination) which makes this seem probable. However, we also know of examples in which cell structures, thought not to contain any nucleic acid, carry 'information', in the sense of specificity which can have an active effect on processes going on in the neighbourhood. For instance, the arrangement of enzymes on the mitochondrial membrane is of this character and, on a larger scale, there is a great deal of evidence for similar operative information in the cortex of egg cells; again one can refer to the growth of cellular organelles, such as the nuclear envelope and stacks of annulate lamellae, where the appearances strongly suggest, though they do not prove, that the existing structural arrangement plays a part in the production of new similar structures in its neighbourhood. There seems to be no good theoretical reason why such information-containing structures should not exist. The main question to be asked about them relates to their capacity for self-replication, that is to say, for information transmission. A very few biologists (e.g. Curtis) think that the cortical information in an egg cell can be transmitted over many generations. The majority of biologists doubt this, but there seems no good reason why such information should not be transmitted through the formation of at least a certain number of replicates as, for instance, in the growth of a mitochondrion or nuclear envelope.

It may be profitable to ask what are the essential requisites for an efficient mechanism of biological information transmission. One most important one would be to make the system independent of the disturbance caused by cell division. DNA has got over this hurdle, since the whole cell division process is obviously carefully organized around the process of DNA replication and the separation of the replicates. Small information carriers, such as small portions of the cortical surface or small organelles within the cytoplasm, could also avoid being disturbed by cell division. If they can operate so as to guide synthetic processes in their neighbourhood so that they build up replicates of themselves, there is no reason why they should not show an indefinite genetic continuity. There may well be cell organelles which have this property (centrosomes, basal bodies of flagellae, etc.?), which, that is to say, have a definite structural organization and cause the appearance of a replica of themselves in their surroundings. However, they would not be information-carrying systems in the full sense unless they could suffer alterations (mutations) and transmit the mutated state. This obviously demands something more than mere capacity for self-replication of a constant structure. It requires, in the first place, that the mutated structure can have a specific effect on the building up of the new replica, that is to say, if the normal structure is A B C, but owing to a mutation a structure A B D appears, it is necessary that A B D should cause another group of A, B, and D sub-units to come together in the proper arrangement. One can envisage a situation in which it might be able to do so provided sub-units D were available, but if the sub-units are themselves produced under the control of some other information system (e.g. chromosomal D N A) the inherent capacity of the A B C to transmit information could not actually be utilized. It seems rather likely that it is for this sort of reason that the D N A system has got such thorough control. It seems possible that cells may be full of entities which could transmit information but only by manipulating sub-units of a complex kind with which the genes do not always provide them. They would be, as it were, capable of manipulating words, but not of building the words they required out of individual letters.

This topic is closely related to what I have called 'template production of non-copies' as seen, for instance, in experimental embryology. In these instances a region of tissue carries a spatial pattern of information which is operative in calling forth the appearance of organs which correspond to, but are not identical with, those from which the information has proceeded. In such systems, however, no question of genetic continuity arises and we will not discuss them further in this place

▸ *Questions about evolution*. There is universal agreement nowadays that the foundations of the theory of evolution are to be found in Darwin's concepts of random variation and the survival of the fittest. However, it is as well to notice that modern orthodox neo-Darwinism, although using the same phrases, has actually changed the meaning of almost all the words, so that what emerges is quite considerably different from what Darwin put forward.

1. 'Random variation'. Darwin was thinking of random phenotypic variation. Neo-Darwinism is thinking of random genotypic variation. Here neo-Darwinism is right, in so far that all the evidence suggests that there are no constraints limiting the freedom of mutation (it's just worth considering whether certain nucleotide sequences might be excluded for some reason, but there is little evidence of it). However, neo-Darwinism seems to me to be wrong in so far as it usually tacitly assumes that randomness of genetic mutation implies randomness of phenotypic variation, and I believe Darwin himself was wrong in so far as he believed that phenotypic variation can be adequately characterized as random – it is doubtful if he actually did believe this.

2. 'Survival of the fittest'. Darwin often argued as though he was really thinking about survival in the sense of an organism living for a long period, and he used

the word fittest to mean 'most able to carry out the ordinary transactions of life', such as running, collecting food, etc. The neo-Darwinist meaning is quite different. For survival they substitute – quite rightly – reproduction; and by fittest they mean 'most effective in contributing gametes to the next generation'. Thus the whole consideration of ability to carry out the ordinary business of life has disappeared from the neo-Darwinist theory, and is replaced entirely by the conception of reproductive efficiency. This in effect reduces Darwinism to a tautology, and leaves for quite separate discussion – which is very rarely provided – why animals should have evolved all sorts of highly adaptive structures to do unlikely things instead of simply being reduced to bags of eggs and sperm like certain parasitic worms. These comments apply to the formal mathematical theory of neo-Darwinism. The most valuable modern evolutionary thought is not to be found in this form of theory but in more generally phrased discussion, and it is the systematization of these views that offers the most profitable subject for consideration.

One may begin by asking what is the major problem on which the theory of evolution attempts to throw light. Again, the answer is quite different nowadays from what it was for Darwin. For him the major problem was to establish the point that species have changed and are derived one from another. Nowadays everyone accepts this, and there is no point in pursuing it. For us the major problem is one which was only a second order issue to Darwin. This is the problem of adaptation. Why do we find animals and plants which have structures and capacities that make them admirably suited to carry out extraordinary living routines in the most unlikely situations, often highly unfavourable for reproduction? A second major problem for evolutionary theory today is to understand how and why living organisms have become divided up into separate taxonomic categories – and although Darwin called his book *The Origin of Species* he said remarkably little about this point.

Modern biological theory holds firmly to the conclusion that the major agencies producing evolutionary change are random mutation and differential reproduction, with minor complications due to migration and hybridization (more important in plants) and essentially no contribution from environmentally directed hereditary variation. The major new information we have accumulated has been concerned with the genetic structure of the populations of organisms in which these processes occur. We know that these populations contain much more genetic variation than would be apparent at first sight. We know that segregation and recombination is an extremely important source of variation in the short term, though ultimately dependent on random mutation in the background. We know further that the variation in a population is 'co-adapted'. However, these points, although of the greatest importance in setting the stage on which evolutionary theories must operate, do not in themselves directly contribute to the solution of the major problems mentioned above. Moreover, as we have seen in the first paragraphs of this section, the mere invocation of reproductive efficiencies and Malthusian parameters leaves the essential problems on one side.

The remaining avenue left to be explored is that opened up by the remark that randomness of genetic variation does not imply randomness of phenotypic variation. We have, in my opinion, to return to Darwin's concern with the nature of phenotypic variation – a subject about which he was always complaining that there was, in his day, a total absence of any understanding. Nowadays not only has the development of genetics given us some insight into genetic variation, but the development of epigenetics is giving us at least a few hints as to the nature of phenotypic variation.

▶ *Some problems for the immediate future of evolution theory*

1. The old problem of the relation between heredity and environment in evolution (what is often, though somewhat incorrectly, spoken of as the problem of Lamarckism) has, I think, largely been cleared up by the recognition that the capacity of an organism to respond to environmental stresses during development is itself a hereditary quality. Further, the demonstration that a combination of this fact with that of developmental canalization leads to the occurrence of a process of genetic assimilation, by which the effect of an 'inheritance of acquired characteristics' can be exactly mimicked, has removed the whole heat out of this ancient discussion. These developments of biological theory have, however, I believe, considerable philosophical implications, since they show how it is possible for characteristics such as mental or perceptive abilities, which originally arose in interaction with the environment, to become in later generations 'inborn' to the extent of being independent of any particular environmental stimuli.

2. There has not yet been sufficient exploration – intellectually, let alone experimentally – of the feedback situation, in which an animal's behaviour largely determines the kind of selection pressure to which it will be subjected. Consideration of this will probably need an elaboration of evolutionary theories involving selection between small, semi-isolated populations or 'demes'. In so far as this process occurs, one could consider a large population of animals as playing a game against nature, making a series of moves in each of which one deme would adopt one type of behaviour and submit itself to the resulting natural selective

pressures, while other demes might choose to operate in a different way. It is in this connection that games theory might have much to offer evolution theory, if indeed it has any theorems to offer anybody; but so far I have not been able to discover what they are.

3. What points of principle, if any, emerge from the fact that we can now begin to draw up evolutionary trees of the changes in amino acid sequences in proteins? So far as I can tell up to date, studies of this kind have no more (and possibly even considerably less) to tell us about the general theory of evolution than do studies of any other phenotypic characters, though they may have something to tell us about the way in which particular proteins carry out their biochemical operations.

4. Recent evolutionary theory has concentrated very largely on questions of the responsiveness of populations to selection pressures by alterations in gene pool. There has been much less development of theories concerned with changes in numbers of a population in competition with other ecological competitors, and in relation to a random or even, as Lewontin calls it, a capricious environment.

One of the more immediate tasks of evolutionary theory is to decide what, if anything, could be meant by 'the fitness of a species (or a population)'. In current technical practice, 'fitness' is a parameter ascribed to individuals of a given genotype in generation N, and measures the relative probability that they will produce offspring which succeed in transmitting that genotype in generation N+1. That is to say, its definition pays attention only to the very short term, and the concept is clearly inadequate in relation to long-term evolution. Thoday has suggested a notion of 'fitness' which is defined in terms of the chances of leaving offspring 10^{6+} generations hence. This suffers from the opposite drawback, that it has no relevance to evolution within geologically reasonable periods, e.g. none of the competing families of Jurassic ammonites survived the Cretaceous. The real question would seem to be: how can we compare, here and now, the effectiveness with which two populations will probably be able to cope with a future which is essentially not completely foreseeable? The most useful concept, in my opinion, would be one in which 'population fitness' would be defined in terms similar to those which one would use to determine the value of a hand at cards, a set-up on a chess board, or, even better, the 'usefulness' of a tool (an adjustable wrench in comparison with a box spanner, say). But how, if at all, can one do this? Perhaps the inventors of games-playing automata have something to offer?

5. One could, in this connection, also ask the same question as was asked in relation to development: What is the basic elementary process of evolution? According to neo-Darwinism it is simply to leave more offspring than your

neighbour. But, as pointed out above, this formulation omits any reference to alterations in any other aspects of phenotype than those concerned with reproduction. A more adequate answer would be something like: To find some phenotypic modification which facilitates leaving more offspring than your neighbour; or, to put it even more generally, to find some way of coping with the situation (I realize that Americans say Britons *cope* with situations while they *change* them).

6. The principle of archetypes in evolution. Evolution is brought about by the natural selection of random variations which occur in the genetic material which – in conjunction with environmental influences – determines the phenotypes of the organisms which have to find some way of earning a living in the natural world. The point towards which attention is being directed in this Note is the fact that the individuals on which natural selection operates are *organisms* – that is to say, their character is not a mere summation of a series of independent processes set going by a number of disconnected genetic factors but is instead the resultant of the interaction (involving all sorts of feedback loops, mutual interference, mutual competition, etc.) of a number of elementary processes for which the individual genetic factors are responsible. Now a major characteristic of such interacting systems is the existence of 'threshold phenomena'. Suppose that we have an organized system subject to variation, and natural selection makes the demand on it that a certain parameter should reach a minimum value p. It will very likely be the case that the first variation which attains this value p actually interlocks with the other elements in the organized system so that, in fact, it attains a value of $p \times n$, where n may be quite a large number.

Consider an excessively simple case. Imagine in a two-dimensional universe an organism which was faced with the necessity to protect its internal contents against the random buffetings of the external world; the organism is to be thought of as having protective elements in the form of three rods of relatively rigid material joined together at the ends by joints around which movement is possible. Now if these three rods are such that the sum of the lengths of the two shorter elements is less than the length of the longer element, they can form a sort of protective arc whose strength can only be increased by increasing the friction at the terminal joints. This will give the whole configuration a certain but quite restricted resistance to deformation. If, however, the length of the two shorter arms is increased so that the whole system can join together to form a triangle, the overall resistance to deformation of the configuration suddenly leaps up by many orders of magnitude – probably by much more than existing circumstances were demanding of it.

I should like to suggest that a major factor in large-scale evolution has been this sort of hitting-the-jackpot. For some relatively trivial local reasons certain arthropods varying indefinitely over a whole range of phenotypes under the influence of natural selection come up with a form involving three legs, two pairs of wings, and respiration by tracheae – it turns out to their surprise (if a phylum can feel such a sentiment!) that this particular pattern of organization opens out the whole range of the insects. Another group comes up with an 'archetype' of eight legs with no proper division between the head and the thorax, etc., and they find they can develop into the whole range of the Arachnida.

There should be – but unfortunately there are not yet – large chapters in the mathematical theory of evolution concerned with such questions as:

1. Under what circumstances does selection for a slight increase in a parameter X in an organized system lead to an increase in X by several orders of magnitude?

2. How can you distinguish between topologically distinct patterns of organization? Consider automobiles. There is the motor bicycle pattern, with one wheel to drive and one to steer and no lateral stability. There is next the three-wheeler, with either one to drive and two to steer or vice versa, and some lateral stability. There is the standard four-wheeler, usually with two to drive and two to steer but possibly with both pairs driving and/or both pairs steering. Then there is the tractor vehicle, which amounts to an infinite number of driving units and I am not sure how many steering units. Is it conceivable that there could be a biological topology which would show why you can get large classes of animals with two, three, or four pairs of limbs, or an indefinite number of pairs as in Myriadopods, but there is no major class with six or seven pairs; or again, you can get symmetries based on two, four, five, and six axes but not apparently on seven? Or is this perhaps merely a contingent accident that no animal phylum during evolution happens to have hit on a seven-ray symmetry? But even without tackling the possibly insoluble problem of why evolution hasn't done what it hasn't done, we certainly cannot be content without some further understanding of how these small-scale processes of natural selection of minor variants in relation to immediate needs have produced a restricted number of basic 'archetypes' – the protozoan, the annelid, the insect, the vertebrate – which are flexible enough to become adapted to almost any system of life yet have sufficient inherent stability to do so without losing their essential character.

3. The notion of an archetype, as described above, is really too 'simpliciste' – for purposes of preliminary exposition. What we really have to do with always involves time, and in the context of evolution, involves it at two (or more) importantly different scales. In the first place, an archetypal form of individual organization (an

insect, an arachnid, etc.) is not simply the conventionally accepted adult configuration – it is the whole epigenetic trajectory leading from the egg to the adult. It is an archetypal chreod. (And here it is worth noticing the difference between an archetypal and the usual kind of epigenetic chreod; both are 'canalized' and protected by threshold-like barriers against disturbances; but in normal chreods the thresholds lie where they do only because natural selection has put them there, by selecting certain values of particular parameters. In an archetypal chreod, on the other hand, the position of the thresholds is fixed by internal necessity (is this what René Thom calls 'stabilité structurelle'?) – you can't have a triangle unless $a + b > c$; of course it is natural selection which picks out one archetypal chreod as something to be exploited, but it has not created it, whereas it *has* created the normal epigenetic chreod out of a situation in which no chreod is logically necessary.)

Secondly, to return to the main argument, an archetype should probably – I am not quite sure of this – be regarded as time-extended on the evolutionary time scale. You don't just get a 'horse archetype', a 'Dipteran archetype', but you get a 'horse family archetype, with inbuilt characteristics of directions in which evolutionary change can easily go'.

THE CONTRIBUTION OF BIOLOGY TO PHILOSOPHY

The difficulty of discussing this is that philosophy in the last quarter of a century has got itself into what is almost universally regarded by scientists as such a fantastic mess that a certain amount of clearing up of strictly philosophical matters has to be done before the subject can even be approached. The most fashionable view of professional philosophical circles is in fact that no form of science – biology or any other – has anything to contribute to philosophy. This is, from a simple historical point of view, such obvious paradoxical nonsense – of Newton and Locke, or Darwin and Bergson – that it can only be defended by claiming that nearly all figures of the past known as philosophers were not really philosophers at all – a conclusion which the more extreme adherents of the modern school have not been afraid to draw. I think that if we are to discuss the matter at all we must plunge in at the deep end, as follows:

▸ *The implications for epistemology of science in general and biology in particular.* The problem of epistemology is the relation between what we (human beings) perceive and the 'real' nature of the entities which appear to exist outside ourselves It is philosophically fashionable to claim that science has nothing to contribute to the understanding of this matter. In my opinion, such plausibility as this claim

possesses rests on a failure to distinguish two different meanings of the word 'perceive'. These two meanings are:
(i) 'Perceive' as a conscious sensation (which we might call to 'experience');
(ii) 'Perceive' as a causally efficacious stimulus to which we react but of which we may be totally unconscious (which one might call to 'react to').
These two meanings are definitely different and the two processes involved can occur separately: for instance, if you fly rapidly up to the city of Cuzco at 12,000 ft. in the Andes, as I recently did, your respiratory rate will increase and your heart will start beating faster although you cannot consciously experience, as a sensation, the change in barometric pressure and oxygen pressure.

Now the epistemological problem of the relation between what we perceive and the nature of the 'real world' obviously depends fundamentally on the *existence* of the phenomenon of experience, if only because the whole question arises within a world of consciousness and could not be formulated or discussed in any manner which did not involve consciousness. Arguments which aim to show that science cannot contribute anything to epistemology are valid only to the extent that they claim that conscious experience can never be shown to be an inevitable resultant of the scientifically describable processes involved in perception (electromagnetic vibrations impinging on the retina, rods and cones, visual nerves, etc., or oscillations of air pressure impinging on the auditory organs, and so on). This is perfectly true as far as it goes, but it does not go very far. If we accept that, in some way that we do not comprehend at all, we do have conscious experiences, then the ground is cleared for science to make the major point which it does wish to make in this connection; namely that we can find out far more about the nature of the 'real world' by paying attention to the way we react to it than by paying attention to the particular character of the conscious experiences we have of it. Our understanding of such things as the atomic nature of material, the existence of electromagnetic vibration, Newtonian, quantum or relativity physics, etc., does not depend upon the particular quality of conscious experience; it could all have been worked out by colour-blind people who could not experientially distinguish red from green. The content of what we can discover about the real world depends on 'reacting to' rather than 'experiencing'.

It is from this point onwards that I think biology has something to contribute. In the first place, it would point out that different organisms differ in the batteries of sense organs they possess, which might be thought capable of giving rise to conscious experiences, and that one might expect that this would make it easier or more difficult for them to learn about certain characteristics of the external real world. Organisms that are totally colour-blind (cannot distinguish between wavelengths of electromagnetic vibration) would presumably – if they could form any conscious picture of the nature of the real world – find it more difficult to come to a picture which involved differences in wavelengths. Organisms which have a very highly developed sense of smell might – if smell really depends on vibrational properties of molecules, as some modern theories suggest – find it more easy to develop an understanding of this aspect of the real world rather than, say, of the bond angles between the atoms in the molecule. Man has, of course, by now very largely – though perhaps not entirely – escaped from these limitations of his natural endowment of sense organs, but only by first exploiting those sense organs to the full and then developing sophisticated apparatus which allowed him to explore, for instance, regions of the electromagnetic spectrum to which he is not physiologically sensitive. Even so, he is having the gravest difficulty in forming conceptual schemes to cope with aspects of the real world which are revealed at a scale of magnitude to which he is not accustomed, such as those in the subatomic or, even more, the intra-nuclear worlds.

A second, more profound contribution of biology would be in connection with the degree to which the involvement of a percipient being, with a particular sensori-neuronal equipment, in any act of perception limits the 'validity' of our world picture. There was a controversy among physicists about this in the not very distant past. Most* of those who believe in the 'Copenhagen' interpretation of quantum phenomena also believe that our scientific knowledge is not nearly as objective as had previously been thought. The argument can start from the conclusion that any means used to observe the position of an elementary particle, such as a ray of light reflected from it, is bound to interfere with its motion, and thus make it impossible simultaneously to determine precisely its velocity. This is one of the forms in which the uncertainty principle is expressed. Heisenberg argues:

> 'If we wish to form a picture of the nature of these elementary particles, we can no longer ignore the physical processes through which we obtain our knowledge of them. . . . Thus, the objective reality of the elementary particles has been strangely dispersed, not into the fog of some new ill-defined or still unexplained conception of reality, but into the transparent clarity of a mathematics that no longer describes the behaviour of the elementary particles but only of our knowledge of this behaviour. The atomic physicist has had to resign

* For lack of time to rewrite it, this point is being made in the form of a quotation from a forthcoming book on the relations between science and painting in this century.

himself to the fact that his science is but a link in the infinite chain of man's argument with nature, *and that it cannot simply speak of nature "in itself"*. Science always presupposes the existence of man and, as Bohr said, we must become conscious of the fact that we are not really observers but also actors on the stage of life. . . . If, starting from the condition of modern science, we try to find out where the bases have started to shift, we get the impression that it would not be too crude an oversimplification to say that *for the first time in the course of history modern man on this earth now confronts himself alone, and that he no longer has partners or opponents*. . . . Even in science, *the object of research is no longer nature itself, but man's investigation of nature.*'

Heisenberg goes on to quote a passage from Eddington, written in his characteristically academic purple style:

'We have found that where science has progressed the farthest, the mind has but regained from nature that which mind has put into nature. We have found a strange footprint on the shores of the unknown. We have devised profound theories, one after another, to account for its origin. At last, we have succeeded in reconstructing the creature that made the footprint. And lo! It is our own.'

Schroedinger voices the opposite view. In one place he remarks rather acidly:

'The argument for whose sake I have inserted this brief report would be reinforced by mentioning the "pulling down of the frontier between observer and observed" which many consider an even more momentous revolution of thought, while to my mind it seems a much over-rated provisional aspect, without profound significance.'

In fact, Schroedinger does not believe that the observer ever has been or should be mixed up with the observed in scientific research; he thinks that it is only because of a temporary failure to reach a full scientific understanding of quantum phenomena that he seems to have got into this anomalous situation at present. His belief is:

'That a moderately satisfying picture of the world has only been reached at the high price of taking ourselves out of the picture, stepping back into the role of a non-observed observer.'

He describes how we have found ourselves faced by

'two most blatant antimonies due to our unawareness of this fact. . . . The first of these antimonies is the astonishment of finding our world picture "colourless, cold, mute". Colour and sound, hot and cold, are our immediate sensations; small wonder that they are lacking in a world model from which we have removed our mental person. The second is our fruitless quest for the place where mind acts on matter or vice versa, so well known from Sir Charles Sherrington's honest search, magnificently expounded in *Man and his Nature.* The material world has only been constructed at the price of taking the self, that is, mind, out of it, removing it; mind is not part of it. It therefore can neither act on it nor be acted on by any of its parts.'

Schroedinger's view was, as I have said, not the fashionable or dominant one among physicists. But it was not far removed from that implicit in the first few lines of the most influential treatise on philosophy of the Thirties, Wittgenstein's *Tractatus Logico-Philosophicus.* This book is mainly concerned with the nature of language, and its relation to the things it refers to. It is written in a series of detached sentences, each of which is numbered in a system which expresses their place within the logical system of the whole argument. And it begins boldly:

'1. The world is everything that is the case.'

(As though we knew what *that* is! The life of most scientists is dedicated to trying to find out what is the case in some minute area of the world's affairs.)

Wittgestein goes on with the appealing point.

'1.1. The world is the totality of facts, not of things'

and proceeds through a number of arguments till he reaches a point where he joins forces with Schroedinger:

'1.2. The world divides into facts. . . .

2. What is the case, the fact, is the existence of atomic facts.

2.01. An atomic fact is a combination of objects (entities, things). . . .

2.0232. Roughly speaking, objects are colourless.'

In this connection, biology would point out that the capacity of an organism to react to factors in the real world has been subject to natural selection and that therefore organisms will in general be able to react to factors which it is important for them to react to. The picture of the real world that they can attain to on the basis of their reactions is therefore a picture of something which is '*really* real' in the sense that it conditions the survival of their species. It is difficult to see what further sense could be attributed to the word 'real' than this.

We would thus come to a view of a kind which would be classified in philosophy as 'critical realism' or something like it. But here 'realism' would not mean that some other external observer – a Martian, or even God–would see external nature in terms of the latest theories of the most advanced physicists of the day – which probably won't last more than ten years in human terms anyway–but rather that a given species forms a picture of the external world of a kind which it pays off for that species to use in its practical business of earning a living and reproducing its kind. And this picture will be extremely complex and, as it were, on very many levels. It is just as much a real part of the picture to see something as a mouse as to see it as a collection of DNA and proteins or a collection of sub-atomic particles.

Among fairly recent metaphysical accounts of epistemology by philosophers, perhaps the nearest to this view has been that of Whitehead in his contention that the basic stuff of the universe consists of 'events' and that everything you can drag out of an event – from a sub-atomic particle to a poetic metaphor – is actually a part of it. What he omitted from his account, however, was the biological point that you have been tailored by natural selection to discover in an event what it is essential for you to know.

▶ *Is ethical value an intrinsically inevitable product of the material universe?* Some philosophers, e.g. Whitehead, have argued that consciousness and value are (a) obvious facts about human existence, (b) of a nature so different to the notions entertained by physical science that no manipulation of these notions could produce them (for instance, as resultants of mere complexity) and have concluded that (c) if we are to be consistent evolutionists we must suppose that some characteristic of the same general order of existence as consciousness and value must exist in all entities (e.g. sticks and stones) even though we have no means of detecting it.

I want to raise a somewhat similar issue but from a more step-by-step, though still wildly speculative, argument.

It is becoming quite easy to put forward a plausible and even convincing argument that the nature of the material world is such that, given a set of circumstances that are bound to exist in a number of places in the cosmos, physico-chemical systems will arise which exhibit the basic properties of 'life' in the sense of: (a) identical duplication, (b) mutation and identical duplication of the mutants (i.e. transmission of information), (c) the elaboration of a penumbra of 'expositions' of the information-carrying elements (i.e. a phenotype). Such a system or systems inevitably give rise to the necessity of evolution by natural selection. It is not too difficult to argue that this will eventually carry the reproducing entities (organisms) to a stage when they are capable of producing a new system of transmitting information which is incomparably more rapid than the simple molecular systems which arose at the beginning of the process. In evolution on earth, this step has been taken by the human species by the systematic transmission of information, not merely through DNA but also by spoken or written words. Now

I have argued (*The Ethical Animal*, 1960) that the particular way this information-transmission system has been developed by man has inextricably connected it with notions of social (usually parental) authority, since it is only the association with socially imposed restraints (essentially somebody saying 'No') that persuades the newborn infant that a modulation of air pressure on its ears does in fact encode information. Further, I have argued that it is the fact that the young human individual makes this fundamentally important evolutionary step – of becoming a member of the network of 'information-transmission through words' – by realizing that words can convey *commands*, which leads him to develop the idea that there are quasi-absolute commands, which he later calls 'ethics', value systems, and so on.

I am prepared to defend the view that this is what has actually happened on the planet Earth. The question I wish to raise now is this: Is it conceivable that there can be a system of information-transmission through symbolic forms (comparable to words) which does not involve any hypertrophy of inter-personal authority (parental or other) into a system comparable to ethics?

So far as I can see, no transmission system can effectively carry information between a transmitter and a recipient unless the recipient accepts the message as meaningful. At the chemical level the problem does not perhaps arise in a very crucial way, although it should be noted that it is no use pushing the DNA of your sperm into an egg unless the egg contains the polymerases capable of transcribing it into a messenger and all the rest of the machinery for turning out a protein according to specification. However, when the message is transmitted in a form which is completely symbolic (such as a word) the question of whether the recipient can (a) realize there is something encoded in this, (b) has a mechanism for decoding it, becomes even more important. I suggest that the more important matter of the two is the former: does this fluctuation of the environment code for a message? (If it does, we'll find some way of discovering what the message is.) In the human species the realization that certain stimuli are message-conveying is tied up with social restrictions on freedom of satisfaction of instinctive needs – and this, as I have argued, gives rise to ethics. The question is: Is there anything else which it could conceivably be tied up to? I should like to suggest that a possible alternative way of evolving an effective system of transmitting information through symbols would have been to tie it up to the 'objectification' of the sensory world. It seems to be the case (best evidence from congenitally blind people to whom sight has been restored in later life) that the sensory world in the earliest stages of life does not contain any defined recognizable persistent objects. It might perhaps be described as a cinema film in which one of Monet's paintings of waterlilies or of Rouen Cathedral would be a still – though even that is not an adequate description because the distinction between the interior and exterior world – the me-and-not-me – is not fully established. The processes of learning 'there is a me and a not-me; there is not only a whirling kaleidoscope of tone values and colours but there are actual things with (more or less) edges' goes on at about the time of life at which, if ever, a new zygote has got to be brought into a symbolic information-transmitting system. Could symbol transmission have been married to objectification rather than to inter-organismic authority?

When we see a definite object – a cat sits on the mat – our nervous system performs an astonishing filtering operation on the frequencies, phases, and directions of propagation of the electromagnetic vibrations impinging on the outer surface of the eye. It turns something which Monet sweated blood to get back to in its raw state into something very highly simplified. However, in the case of the human species, the simplification is into a visual schema which is quite different to anything that the human species has a transmitting mechanism for broadcasting. We have settled on emitting waves of air pressure to carry information about schematized data, most of which have been distilled out of excessively complicated electromagnetic vibrations in the visual range. (We later went in for visual transmission through alphabets, etc., but this is a secondary complication.) One can theoretically imagine a hypothetical set of organisms which would communicate with one another, not by modulating air pressure but by emitting electromagnetic signals. The baby in its nursery could point and, instead of saying 'The cat is on the mat', transmit a schematic outline of the scene – which would be a highly abstract distillation of what was actually impinging on its retina.

Now the attainment of the stage of 'objectification' – seeing that the world contains certain things with outlines round them, that you can actually pick up and move from place to place and so on – is an achievement that natural selection would certainly have brought about quite independently of any possibility of transmitting information symbolically. It is a feat necessarily achievable by birds that weave firm nests out of a lot of grass stems and similarly for a whole lot of animals that cannot convey the information symbolically. I should like to throw out the suggestion that if, in evolution, you happen to light on a method of transmitting information in the forms of symbols and the symbols are in a different sensory mode from that in which you learn the process of objectification, then you are likely to finish up with information transmission inextricably mixed up with inter-organismic authority, which will lead you into something like a system of ethical values. If, on the other hand, you could transmit your information in the same sensory mode as that in which you learn to simplify the impinging external stimuli to the point of identifying objects, then you might get an evolutionarily effective information-transmitting system which did not involve any notion of ethical value. However, the penalty for this might be that you would have to live in a world in which conventionalized symbolic forms had the same quasi-absolute character as our ideas of right and wrong. The proposition that a circular disc seen from the side looks oval would presumably to such people not seem morally offensive but just intellectually as crazy as some of the quantum-mechanical notions that things don't have definite locations or speeds seem to us.

Comments by René Thom*

Finality in biology. One knows that in classical mechanics the evolution of a system can be described either by local differential equations, such as the equation of Hamilton, $P = \partial H/\partial q$, $Q = -\partial H/\partial p$, or by a global principle of variation, such as the principle of least action of Maupertius. There is equivalence between the two statements, even though one exhibits a determinist point of view, and the other a finalist point of view.

It is probable that in biology every epigenetic or homeostatic process is susceptible in the same way of a double interpretation, determinist and finalist. One must not forget that the essential unit entity in biology is not the isolated individual, but is the continuous configuration in space time which connects a parent individual with its descendants (or more generally the union of such configurations relative to species which have among themselves functional interactions such as predation, partnership in fertilization, etc.). I think that to every adaptive process one could associate a function S of the 'biological state', which would measure in some way the local 'complexity' (or negentropy) relative to the process considered. Between two times t_0, t_1 (for example, parent at the age A and its descendant at the same age A), the continuous configuration would evolve in such a manner as to realize a minimum for the global complexity, $\delta\int_{t_0}^{t_1} S\,\partial t = 0$. In this way there will be realized the minimum complexity, that is to say the most economical adaptation of the process being considered. Natural selection is doubtless only one of the factors in this evolution; I believe, personally, that internal mechanisms – effectively Lamarckian – operate in the same sense.

In complete distinction to the case of classical mechanics, one cannot hope to

* Roughly translated by C.H.W.

see this evolution becoming realized in a manner which is continuous and differentiable at every point in the individual; the evolution is the result of a great number of local variations of a stochastic character of which only the general result is guided by the variational principle. Because of this, discontinuous variations are not excluded.

Although, in our observations of living beings, the teleological aspect of organs and behaviour strikes us immediately (by analogy with that which we are, and our own behaviours as human animals), the deterministic and mechanistic aspect tends to escape us because it has to do with an effect of very long duration, of a statistical character, 'solidaire de l'évolution', in which the decisive factors (influence of metabolism on chance mutation) are probably extremely tenuous.

▸ *The notion of feedback*. Like all notions taken from human technology, the notion of feedback cannot properly be invoked to explain the stability of biological processes; at most it can serve to illustrate the oscillatory character of the passage towards equilibrium (which one meets in mathematics in the process of successive approximations). The only notion which is mathematically and mechanically acceptable is that of 'structural stability'. In fact structural stability is a natural and *indispensable* attribute of every identifiable form (a form which is unstable in its structure ceases to be a form; it is 'informal'). I add, however, that the stability of biological forms presents certain peculiar aspects; it is not absolute, because perturbations which are apparently very feeble (such as the introduction of a molecule of DNA into the metabolism) are able to destroy it; the stability is particularly marked in certain perturbations, those of which one says, for this reason, that they are in the repertoire of the 'genetic patrimony' of the species.

▸ *Homeostasis and homeorhesis*. The distinction between homeostasis and homeorhesis is familiar to the technician of differential equations; homeostasis signifies that the point representing the state of the system rests in the neighbourhood of a position of stable equlibrium in the phase space; homeorhesis signifies that the representative point finds itself in the neighbourhood of an invariant ensemble of trajectories K, which is an attractor (or at least a 'centre') for the trajectories of a neighbourhood. In the simplest case, the ensemble K is a closed trajectory. One knows that then the local study of the situation is not functionally different from that of homeostasis. It is quite different if the topological structure of the invariant ensemble K, is more complicated, for instance if it is a variety such as a torus. If, following a deformation of the dynamic system, this attractive ensemble comes to be destroyed, it can give place to a great variety of situations: the ensemble K can be replaced by a finite ensemble of attractors of inferior dimensions (even of isolated points). There is then a degeneration of a homeorhetic situation into a homeorhetic situation of lower dimension, possibly even into a homeostatic situation. In all cases, this degeneration becomes expressed in morphogenesis by a generalized catastrophe of a katabolic character. Such a catastrophe is, if not irreversible, at least very difficult to reverse. One comes across this situation (probably) in the phenomenon of induction in embryology (where the situation at the end of the catastrophe is homeorhetic in a lower dimension) or at the death of the living being where the final situation is evidently homeostatic.

This last example shows clearly that the only important notion in biology is that of homeorhesis; there is homeostasis only after the arrest of metabolism, that is to say the virtual death of the living being. From this point of view the synthesis of living matter from inert matter requires a veritable generalized catastrophe of an anabolic kind: the formation of a homeorhetic situation from the basis of a homeostatic situation requires the realization of an infinity of local syntheses brought about in a finite time following a well-defined spatio-temporal scheme. Unless it merely makes a direct simulation of vital metabolism, one may doubt whether traditional chemistry will ever succeed in this task.

▶ *Molecular biology and genetics*. Genes are the veritable 'deus ex machina' of modern biology; although their role is probably not overestimated in this, still it is necessary to put them back into the fundamental perspective which sees life as a dynamic phenomenon. Their raison d'être is evidently their role as regulators of metabolism or of epigenesis; let us give here a scheme, of a simple-minded geometry, which illustrates this role. Suppose that the cytoplasm, in order to carry out the vital functions in the best possible manner, ought to exhibit a local metabolic regime (o) (optimal regime); then suppose that in consequence of perturbations of the external environment, the local metabolic regime is deformed into a sub-optimal regime (s); this situation cannot continue for long without grave consequences for the cell or its reproduction; the 'regulation' or return to the optimal state will be manifested geometrically as follows: There will appear, at a given time t, a very small region of the regime (o) surrounded by a shock wave (W), which will at first be punctate, which separates this region from the surrounding regime (s). In the following periods this shock wave W will develop, enlarge, and fill the whole of the cell so that after a certain time the whole cell will find itself again in the state (o). This is obviously an idealized scheme, much too simple, which is never more than partly realized in nature; moreover in cells with nuclei the role of the variable shock waves is probably taken by the nuclear membrane which is actually relatively fixed; further, nothing prevents us imagining that the regimes (o) and (s) could be deformed continuously one into the other, so that the shock wave (W) could have a free edge, which will be at every moment a curve (following the phenomenon of Riemann-Hugoniot). One could from that imagine that the 'point of initiation' of the regime (o), and the curved edge of the shock wave (W), are connected with specialized molecular structures: the curve will be the bacterial chromosome (or portion of it), and the point of initiation is the 'operator' of Jacob-Monod. The surface W is the region swept out by the filaments of messenger RNA which become detached from the chromosome; very probably this region has the character of a surface only in the neighbourhood of the chromosome; further out it takes the form of a sector of a three-dimensional fan; in this sector, the filaments of messenger RNA capture the ribosomes, where there takes place the synthesis of the enzyme proteins destined to facilitate the transition from the regime (s) to regime (o). When the optimal regime (o) has invaded the whole space, the activity of the chromosomes becomes extinguished.

One might ask why in the above scheme the singularities of metabolism are necessarily connected with specific molecular structures; we probably touch here on one of the crucial problems of vital dynamics. Without wishing to pretend to provide a definite answer, here is a conceivable mechanism: a macromolecule (or more generally a macromolecular structure) exercises on the surrounding metabolism a catalytic or enzymatic action which moreover varies with the position of the point considered in relation to the structure. Geometrically, such an action will be defined by a vector field in the local biochemical space, which will become singular (infinite) on the molecule itself; given a local state of metabolism one can suppose that thermal agitation will form, more or less at random, molecular structures, which are in general labile; but if these molecular structures exert an enzymatic effect on the surrounding metabolism which tends to reduce the local production of entropy, their stability will for that reason be increased (by application of the principle of Le Chatelier); in the situation in which the local metabolism is relatively homogeneous, these structures could be formed in a statistically homogeneous manner within the space; but if the local metabolism presents discontinuity, or a singularity where there is a massive production of entropy, then one would expect to see the formation of molecular structures M geometrically connected to the singularity, relatively stable, and characteristic of the metabolic nature of the singularity (that is to say in such a way as to reduce the formation of entropy). Further, these molecular structures, once constructed, could have an internal stability which would allow them to survive the singularities in metabolic conditions which had given rise to them. When, later, metabolic conditions similar to those of the singularity S return, then the presence of the existing molecular structures M would have an enzymatic effect favouring a return to the singular metabolic state S. There would then be, as an overall result depending on the enzymatic coupling between metabolism and molecular structure, an effect of 'facilitation' or of 'memory' : the probability of re-obtaining the singularity S would be increased after it had been realized a large number of times, because the structures M catalysing the presence of S have been able to appear and to persist. A mechanism of this kind makes it possible to imagine how a mechanism of the 'feedback' type could be built up naturally. Let us give a simplified model : let u be a scalar variable of metabolism (for example the concentration of a given metabolite) ; one would suppose that after a certain disturbance u tends towards an equilibrium optimum value $u=c$, and does so by a process of successive oscillations in which the function $u(t)$ of the time t presents consecutive maxima and minima ; one might further suppose that this oscillation was connected to a spatial circulation of the metabolite under consideration, so that the maxima are realized in a region A of the cell, the minima in some other region B different from A ; if the amplitude of the oscillations attains a threshold f, one can suppose that the region A will be the site of molecular syntheses favouring the passage of the derivative du/dt from a positive value to a negative one. In the same way in the region B one could have a spontaneous synthesis of molecules favouring the inverse effect. The presence of these molecules will tend to favour the global stability of the function (u), which, whatever the amplitude of the disturbances, can thereafter only oscillate between $c-f$ and $c+f$.

This model exhibits a very interesting phenomenon : the production, from a situation of equilibrium with differentiable oscillations (a 'centre' in differential geometry), of a 'canalized' situation with reflecting walls where the process has a derivative (speed) which is discontinuous ; in the theories of regulation of Jacob-Monod, such a transformation would be interpreted as a mutation bringing about the change from the 'constitutive state' to the inducible or repressive state ; in the model above, such a mutation is attributed to the presence of specific particles, and ought to be dominant in diploids ; if there exist dominant constitutive mutants (the O^c mutations) it is because the mechanism envisaged is itself, by origin and necessity, a corrective mechanism (for example the synthesis of an enzyme) appealing to the presence of specific particles, and because the mutation has its effect on the coupling between the effects of two types of particles. One can hardly see a limit to the complexity of the cascades of couplings and of controls which can intervene in such systems. To attribute each possible coupling to a gene, as is usually done, does no more than push the problem back a stage; since the phenotypic effect of genes of order k will be controlled by genes of order (k+1); one finds oneself on an infinite regress: Quis custodiet ipsos custodes?

It seems inevitable, in these conditions, to postulate the existence of metabolic factors which have some genetic properties; I personally believe that there is a whole continuous range of genetic entities between the classical chromosomal gene (with Mendelian behaviour in metazoa) and factors which choose between metabolic regimes, with properties more or less labile and fluctuating. By the phenomenon of 'facilitation' or of 'memory' described above, every type of metabolic regime will tend to support the synthesis of characteristic particles which reinforce the probability of the appearance of the regime. These particles have from this fact the apparent property of automultiplication; one could consider them as virus, if they were capable of forming themselves into stable virus particles outside the cell. The thermodynamic competition between two possible metabolic regimes could be interpreted as a struggle between two populations of 'plasmagene-particles'; from this point of view, natural selection operates just as much in the interior of the metabolism of each individual as it does between the individuals and between species. The chromosomal character of a genetic factor is sometimes reversible (example, temperate phages), and one can very well imagine, from this fact, the progressive incorporation of genetic factors into the chromosomal complement. A description of this hypothetical mechanism could only attain a little precision if one had to begin with a good comprehension of the early historical origins and the dynamic justification of the chromosome and of nuclear material in general.

▸ *Morphogenesis and genetics.* The role of the nuclear material in morphogenesis is a ticklish subject to state precisely. The fundamental process of morphogenesis, of cellular differentiation in particular, is, dynamically, a process of 'catastrophe': an attractor of the metabolic regime – initially homogeneous all over a domain D, undergoes a topological deformation which transforms it into a finite (or infinite) number of new attractors; the competitions between these attractors give rise within D to shock-waves,* which after a certain time rigidify into the boundaries of organs. In the phenomenon of non-living nature, such as changes of phase

* As I (very tentatively) understand it, the word 'shock-wave' is used for the boundaries between regions of multidimensional phase space, in which one can represent the totality of biochemical processes by a field of vectors, which in each separate region are orientated so that they concentrate on an 'attractor'.

(fusion, condensation, etc.) these processes go on locally in a very indetermined manner, and thermodynamics only prescribes the global evolution of the system; in embryology, equally, the beginning of a catastrophe – which consists of a great number of phenomena so small that they are inobservable – are probably largely without assignable causation ('aléatoire'), but, within a short time, the structures condense and simplify following well-defined schemes (the chreods of Waddington). One conceives that the genotype, by prescribing the presence or absence of this or that enzyme which favours one of the competing regimes to the detriment of the others, could modify in an important way the final state of the catastrophe; but a mathematical model for a situation of this type would be very difficult to work out. The genotype also operates in controlling mitosis (its rate and its direction), that is to say the 'polarization' of the tissue; one is dealing here with an oriented thermodynamics totally different from classical thermodynamics, and here again, a mathematical theory would have difficulty to go beyond a qualitative stage. One can then summarize the role of DNA in epigenesis in stating that it 'guides' the catastrophes of morphogenesis; but it would be going too far to say, as is often said, that it brings them about.

The contemporary successes of molecular biology have perhaps produced some illusions among biologists; the present-day impasse in the physics of elementary particles ought to have made them a little more circumspect; when one goes towards the infinitely small, it is the rule that phenomena become more complicated, and, very likely, biology will give us a new example of this. It is highly possible, for example, that the epigenesis of a mammal may be in its broad outlines easier to analyse than that of a bacteriophage; in the former, in effect, statistical simplifications are to be expected, while in the latter it is a matter of chemical affinities of extraordinary specificity and of improbable refinement. One should consider, for example, that there exists no formal proof of the topological continuity between a virus particle infecting a bacterium and its descendants formed in the lysed bacterium. Even if one admits that this continuity is probable it is undeniable – because of the facts of genetic recombination – that it involves molecular transactions of great complexity. One is a long way from the topological simplicity of reproduction by eggs or by buds in metazoa.

▸ *Chance and mutation*. That mutations have a character which is strictly random (if such an expression has sense) is one of the dogmas of present-day biology. It seems to me, however, that this dogma contradicts the physical principle of action and reaction; of two mutations m and m′ that can occur, that which has the happier effect on metabolism (i.e. that which minimizes the production of entropy) ought to have a slightly greater chance of occurrence. In the classical diagram of the theory of information

Source → Channel → Receiver

it is clear that the source has an effect on the receiver; therefore the receiver has a reciprocal effect on the source; this inverse effect is usually unobservable, because the energy belonging to the source is very great in comparison with the energy of interaction. That is certainly not the case for nucleic acid, of which the inherent binding energy is certainly feeble in comparison with the energies which come into play in metabolism. The objection may be raised that here the receiver is an 'open' system: it is conceivable that the DNA exerts a directive action on the metabolism, without bringing into play any really important energy of interaction; a situation like that of the 'aiguilleur' (the man who controls what are known in English as the 'points' or in American as the 'switches'), who determines the direction in which the train goes, while the inverse effect of the train on the aiguilleur is nil. But, like all examples taken from human technology, the comparison is pretty specious; it does not apply, in any case, except in a state of null metabolism; how could it happen that the switch man should manœuvre his switches from the train while it was in full motion? Now very probably the majority of spontaneous mutations occur in interphase, in full metabolic activity. The breakages and translocations of chromosomes seen in metaphase are only the visible result of metabolic accidents which have occurred before anaphase and which have disarranged the course of the anaphase catastrophe. People often speak of errors in the process of the duplication of DNA; I have difficulty in adhering to the present-day belief, which suggests that a change in a pair of nucleotides (or the introduction of a mutagenic extra pair) is sufficient to render a gene non-functional; it is, it seems to me, to repeat the error of the morphologists who believed that the destruction of one neuron in the brain would prevent thought. However, even if one adopts this anthropomorphic point of view of a mechanism of reading the DNA, which would be upset by a simple mistake ('lapsus'), could not one – pushing the anthropomorphism to the limit – admit that these mistakes are orientated, as suggested by Freudian theory, by the needs and the 'unconscious' drives of the ambient metabolism? It seems to me difficult to avoid the conclusion that the metabolism has an effect, doubtless very feeble, but in the long run perhaps dominant, on the statistical frequency of different mutations. It is this long-term effect which could explain the variation principle of minimum entropy, and the finalistic appearance of biological processes of which we spoke at the beginning.

▸ *Evolution.* Let us recall the objection – very well founded – which the finalists make to a mechanistic theory of evolution: how can we explain, if evolution is governed by the chance of mutations which are controlled only by natural selection, that it has been able to produce more and more complex structures, culminating in man and the extraordinary exploits of human intelligence? I believe that there is scarcely more than one way of escaping from this objection, and that will perhaps be accused of idealism. When the mathematician Hermite wrote to Stieltjes: 'The numbers seem to me to exist outside ourselves and impose themselves on us with the same necessity, the same inevitableness, as sodium and potassium', he did not go far enough, to my taste. If sodium or potassium exist it is because there exists a corresponding formal structure which assures the stability of the atoms Na or K; one can explain this formal structure in terms of quantum mechanics for a simple entity, such as the molecule of hydrogen; it is much less well known in the case of the atoms of Na or K, but there is no reason to doubt its existence. I believe, similarly, that in biology there exist formal structures, in fact geometric entities, which prescribe the only forms which a dynamic system of auto-reproduction can present in a given environment. To employ a more geometric language, let us suppose that the evolution of a biochemical soup in a given milieu can be described by the trajectories of a field of vectors in an appropriate function space; let us designate as F_t the operator which consists in gliding along the trajectories during a lapse of time t; if one takes at the beginning an initial datum defined by a form (s) this form will be auto-reproductive, if, at the end of the time t, the form $F_t(s)$ which is produced is the disjoint topological sum $S_1 \cup S_2$ of two forms isomorphic with s; one will then be led to investigate, for every value of (t) the 'spectrum' of the operator of duplication F_t, that is to say the assemblage of forms s such that $F_t(s) = S_1 \cup S_2$. This spectrum is probably discrete, and it is not impossible that one could define, on its proper elements, operations such as topological sum, topological product, etc. That is to say, one could define within an abstract functional space the domain of existence and of stability of each of these proper forms (as functions of external factors); one would thus obtain a 'paleogenic' (paleontological?) map analysed into chreods, formally analogous to the epigenetic landscape of Waddington; thus evolution is nothing other than the propagation of an immense wavefront across this paleogenic space; exactly as in the development of the embryo, phenomena of induction, of 'catastrophes', of regressions . . . are to be expected, and for the same formal reasons; the motive phenomenon is probably the attraction of forms: every proper form (I would say an 'archetype' if the word did not have a finalistic connotation) aspires towards existence and attracts the wavefront of existing beings when that has reached neighbouring topological forms; there is competition between the attractors, and one could speak of the 'malignity' of a form as its attractor power on neighbouring forms; from this point of view one would be tempted to attribute the apparent arrest of evolution today to the excessive (too malign) force of the human attractor. Of all the living forms theoretically possible only an infinite minority are touched by the wavefront and come through to existence.

The global finalists (à la Teilhard de Chardin) could advance the analogy between evolution and the epigenesis of the embryo in order to claim that, just as the embryo develops in conformity with a plan, the wave of existence deploys itself in structural space in conformity with a Plan, imminent and pre-established; this is to forget an essential difference; the development of an embryo is reproducible and from that fact is an object of science; the wave of evolution is not. To affirm that a phenomenon which is unique and unreproducible proceeds according to a plan is a type specimen of an affirmation which is gratuitous and idle.

I do not overlook the fact that one could advance the same reproach against the scheme of the paleogenic map; if we are in principle dealing with non-reproducible phenomena the map could be no more than a gratuitous intellectual scheme. To that I would reply that the only possible merit of the scheme is to draw attention to local analogies, and it by no means excludes that one could (in particular on lower beings with great reproductive rate) experimentally bring about local evolutionary changes in directions decided on in advance, cf. for example Waddington's experiments on Drosophila. Going to the depths of the problem, one should note that every point in an embryo is at a small distance from a cellular nucleus, which contains, in virtual form at least, all the necessary information for the local realization of the plan of the individual; in the wave of evolution, the only factors common to all the living beings are the elementary biochemical constituents DNA, RNA, proteins, genetic code, etc., which do not seem to constitute sufficiently complex structures to be the support for a global plan for life in general. That is why, until we get further information, I prefer to think that evolutionary development probably proceeds conformably with a purely local determinism.

Cause and Effect in Biology

Reprinted from Science, 134, *1501 (1961)*

Ernst Mayr
Harvard University

Being a practising biologist I feel that I cannot attempt the kind of analysis of cause and effect in biological phenomena that a logician would undertake. I would instead like to concentrate on the special difficulties presented by the classical concept of causality in biology. From the first attempts to achieve a unitary concept of cause, the student of causality has been bedevilled by these difficulties. Descartes's grossly mechanistic interpretation of life, and the logical extreme to which his ideas were carried by Holbach and de la Mettrie, inevitably provoked a reaction leading to vitalistic theories which have been in vogue, off and on, to the present day. I have only to mention names like Driesch (entelechy), Bergson (élan vital), and Lecomte du Noüy among the more prominent authors of the recent past. Though these authors may differ in particulars, they all agree in claiming that living beings and life processes cannot be causally explained in terms of physical and chemical phenomena. It is our task to ask whether this assertion is justified, and, if we answer this question with 'no', to determine the source of the misunderstanding.

Causality, no matter how it is defined in terms of logic, is believed to contain three elements: (i) an explanation of past events ('a posteriori causality'); (ii) prediction of future events; and (iii) interpretation of teleological – that is, 'goal-directed' – phenomena.

The three aspects of causality (explanation, prediction and teleology) must be the cardinal points in any discussion of causality and were quite rightly singled out as such by Nagel [1]. Biology can make a significant contribution to all three of them. But before I can discuss this contribution in detail I must say a few words about biology as a science.

▸ *Biology*. The word *biology* suggests a uniform and unified science. Yet recent developments have made it increasingly clear that biology is a most complex area – indeed, that the word *biology* is a label for two largely separate fields which differ greatly in method: *Fragestellung* and basic concepts. As soon as one goes beyond the level of purely descriptive structural biology one finds two very different areas, which may be designated functional biology and evolutionary biology. To be sure, the two fields have many points of contact and overlap. Any biologist working in one of these fields must have a knowledge and appreciation of the other field if he wants to avoid the label of a narrow-minded specialist. Yet in his own research he will be occupied with problems of either one or the other field. We cannot discuss cause and effect in biology without first having characterized these two fields.

▸ *Functional biology*. The functional biologist is vitally concerned with the operation and interaction of structural elements, from molecules up to organs and whole individuals. His ever-repeated question is 'How?' How does something operate, how does it function? The functional anatomist who studies an articulation shares this method and approach with the molecular biologist who studies the function of a DNA molecule in the transfer of genetic information. The functional biologist attempts to isolate the particular component he studies, and in any given study he usually deals with a single individual, a single organ, a single cell, or a single part of a cell. He attempts to eliminate or control all variables, and he repeats his experiments under constant or varying conditions until he believes he has clarified the function of the element he studies. The chief techniques of the functional biologist is the experiment, and his approach is essentially the same as that of the physicist and the chemist. Indeed, by isolating the studied phenomenon sufficiently from the complexities of the organism, he may achieve the ideal of a purely physical or chemical experiment. In spite of certain limitations of this method, one must agree with the functional biologist that such a simplified approach is an absolute necessity for achieving his particular objectives. The spectacular success of biochemical and biophysical research justifies this direct, although distinctly simplistic, approach.

▸ *Evolutionary biology*. The evolutionary biologist differs in his method and in the problems in which he is interested. His basic question is 'Why?' When we say 'why' we must always be aware of the ambiguity of this term. It may mean 'how come?', but it may also mean the finalistic 'what for?' It is obvious that the evolutionist has in mind the historical 'how come?' when he asks 'why?' Every organism, whether individual or species, is the product of a long history, a history which indeed dates back more than 2000 million years. As Max Delbrück [2] has said. 'a mature physicist, acquainting himself for the first time with the problems of biology, is puzzled by the circumstance that there are no 'absolute phenomena" in biology. Everything is time-bound and space-bound. The animal or plant or micro-organism he is working with is but a link in an evolutionary chain of changing forms, none of which has any permanent validity.' There is hardly any structure or function in an organism that can be fully understood unless it is studied against this historical background. To find the causes for the existing characteristics, and particularly adaptations, of organisms is the main preoccupation of the evolutionary biologist. He is impressed by the enormous diversity of the organic world. He wants to know the reasons for this diversity as well as the pathway by which it has been achieved. He studies the forces that bring about changes in faunas and floras (as in part documented by paleontology), and he studies the steps by which have evolved the miraculous adaptations so characteristic of every aspect of the organic world.

We can use the language of information theory to attempt still another characterization of these two fields of biology. The functional biologist deals with all aspects of the decoding of the programmed information contained in the DNA code of the fertilized zygote. The evolutionary biologist, on the other hand, is interested in the history of these codes of information and in the laws that control the changes of these codes from generation to generation. In other words, he is interested in the causes of these changes.

Many of the old arguments of biological philosophy can be stated far more precisely in terms of these genetic codes. For instance, as Schmalhausen, in Russia, and I have pointed out independently, the inheritance of acquired characteristics becomes quite unthinkable when applied to the model of the transfer of genetic information from a peripheral phenotype to the DNA of the germ cells.

But let us not have an erroneous concept of these codes. It is characteristic of these genetic codes that the programming is only in part rigid. Such phenomena as learning, memory, nongenetic structural modification, and regeneration show how 'open' these programmes are. Yet even here there is great specificity, for instance with respect to what can be 'learned', at what stage in the life cycle 'learning' takes place, and how long a memory engram is retained. The programme then, may be in part quite unspecific, and yet the range of possible variation is itself included in the specifications of the code. The codes, therefore, are in some respects highly specific; in other respects they merely specify 'reaction norms' or general capacities and potentialities.

Let me illustrate this duality of codes by the difference between two kinds of birds with respect to 'species recognition'. The young cowbird is raised by foster parents – let us say in the nest of a song sparrow or warbler. As soon as it becomes independent of its foster parents it seeks the company of other young cowbirds, even though it has never seen a cowbird before. In contrast, after hatching from the egg a young goose will accept as its parent the first moving (and preferably also calling) object it can follow and become 'imprinted' to. What is programmed is in one case a definite 'gestalt', in the other merely the capacity to become imprinted to a 'gestalt'. Similar differences in the specificity of the inherited programme are universal throughout the organic world.

Let us now get back to our main topic and ask: Is *cause* the same thing in functional and evolutionary biology?

Max Delbrück, again, has reminded us [2] that as recently as 1870 Helmholtz postulated 'that the behaviour of living cells should be accountable in terms of motions of molecules acting under certain fixed force laws'. Now, says Delbrück correctly, we cannot even account for the behaviour of a single hydrogen atom. As he also says 'any living cell carries with it the experiences of a billion years of experimentation by its ancestors'.

Let me illustrate the difficulties of the concept of causality in biology by an example. Let us ask: What is the cause of bird migration? Or more specifically: Why did the warbler on my summer place in New Hampshire start his southward migration on the night of 25th August?

I can list four equally legitimate causes for this migration.

1. *An ecological cause.* The warbler, being an insect eater, must migrate, because it would starve to death if it should try to winter in New Hampshire.

2. *A genetic cause.* The warbler has acquired a genetic constitution in the course of the evolutionary history of its species which induces it to respond appropriately to the proper stumuli from the environment. On the other hand, the screech owl, nesting right next to it, lacks this constitution and does not respond to these stimuli. As a result, it is sedentary.

3. *An intrinsic physiological cause.* The warbler flew south because its migration is tied in with photoperiodicity. It responds to the decrease in day length and is ready to migrate as soon as the number of hours of daylight have dropped below a certain level.

4. *An extrinsic physiological cause.* Finally, the warbler migrated on 25th August because a cold air mass, with northerly winds, passed over our area on that day. The sudden drop in temperature and the associated weather conditions affected the bird, already in a general physiological readiness for migration, so that it actually took off on that particular day.

Now, if we look over the four causations of the migration of this bird once more we can readily see that there is an immediate set of causes of the migration, consisting of the physiological condition of the bird interacting with photoperiodicity and drop in temperature. We might call these the *proximate* causes of migration. The other two causes, the lack of food during winter and the genetic disposition of the bird, are the *ultimate* causes. These are causes that have a history and that have been incorporated into the system through many thousands of generations of natural selection. It is evident that the functional biologist would be concerned with analysis of the proximate causes, while the evolutionary biologist would be concerned with analysis of the ultimate causes. This is the case with almost any biological phenomenon we might want to study. There is always a proximate set of causes and an ultimate set of causes: both have to be explained and interpreted for a complete understanding of the given phenomenon.

Still another way to express these differences would be to say that proximate causes govern the responses of the individual (and his organs) to immediate factors of the environment while ultimate causes are responsible for the evolution of the particular D N A code of information with which every individual of every species is endowed. The logician will, presumably, be little concerned with these distinctions. Yet the biologist knows that many heated arguments about the 'cause' of a certain biological phenomenon could have been avoided if the two opponents had realized that one of them was concerned with proximate and the other with ultimate causes. I might illustrate this by a quotation from Loeb [3]: 'The earlier writers explained the growth of the legs in the tadpole of the frog or toad as a case of adaptation to life on land. We know through Gudernatsch that the growth of the legs can be produced at any time, even in the youngest tadpole, which is unable to live on land, by feeding the animal with the thyroid gland.'

Let us now get back to the definition of 'cause' in formal philosophy and see how it fits with the usual explanatory 'cause' of functional and evolutionary biology. We might, for instance, define cause as 'a nonsufficient condition without which an event would not have happened', or as 'a member of a set of jointly sufficient reasons without which the event would not happen' (after Scriven [4]). Definitions such as these describe causal relations quite adequately in certain branches of biology, particularly in those which deal with chemical and physical unit phenomena. In a strictly formal sense they are also applicable to more complex phenomena, and yet they seem to have little operational value in those branches of biology that deal with complex systems. I doubt that there is a scientist who would question the ultimate causality of all biological phenomena – that is, that a causal explanation can be given for past biological events. Yet such an explanation will often have to be so unspecific and so purely formal that its explanatory value can certainly be challenged. In dealing with a complex system an explanation can hardly be considered very illuminating that states: 'Phenomenon *A* is caused by a complex set of interacting factors, one of which is *b*.' Yet often this is about all one can say. We will have to come back to this difficulty in connection with the problem of prediction. However, let us first consider the problem of teleology.

▸ *Teleology.* No discussion of causality is complete which does not come to grips with the problem of teleology. This problem had its beginning with Aristotle's classification of causes, one of the categories being the 'final' causes. This category is based on the observation of the orderly and purposive development of the individual from the egg to the 'final' stage of the adult, and of the development of the whole world from its beginnings (chaos?) to its present order. Final cause has been defined as 'the cause responsible for the orderly reaching of a preconceived ultimate goal'. All goal-seeking behaviour has been classified as 'teleological,' but so have many other phenomena that are not necessarily goal-seeking in nature.

Aristotelian scholars have rightly emphasized that Aristotle – by training and interest – was first and foremost a biologist, and that it was his preoccupation with biological phenomena which dominated his ideas on causes and induced him to postulate final causes in addition to the material, formal, and efficient causes. Thinkers from Aristotle to the present have been challenged by the apparent contradiction between a mechanistic interpretation of natural processes and the seemingly purposive sequence of events in organic growth, in reproduction, and in animal behaviour. Such a rational thinker as Bernard [5] has stated the paradox in these words:

> There is, so to speak, a pre-established design of each being and of each organ of such a kind that each phenomenon by itself depends upon the general forces of nature, but when taken in connection with the others it seems directed by some invisible guide on the road it follows and led to the place it occupies.
>
> We admit that the life phenomena are attached to physicochemical manifestations, but it is true that the essential is not explained thereby; for no fortuitous coming together of physicochemical phenomena constructs each organism after a plan and a fixed design (which are foreseen in advance) and arouses the admirable subordination and harmonious agreement of the acts of life. . . . Determinism can never be [anything] but physiochemical determinism
>
> The vital force and life belong to the metaphysical world.

What is the *x*, this seemingly purposive agent, this 'vital force', in organic phenomena? It is only in our lifetime that explanations have been advanced which deal adequately with this paradox.

The many dualistic, finalistic, and vitalistic philosophies of the past merely replaced the unknown *x* by a different unknown, *y* or *z*, for calling an unknown factor *entelechia* or *élan vital* is not an explanation. I shall not waste time showing how wrong most of these past attempts were. Even though some of the underlying observations of these conceptual schemes are quite correct, the supernaturalistic conclusions drawn from these observations are altogether misleading.

Where, then, is it legitimate to speak of purpose and purposiveness in nature, and where is it not? To this question we can now give a firm and unambiguous answer. An individual who – to use the language of the computer – has been 'programmed' can act purposefully. Historical processes, however, can *not* act purposefully. A bird that starts its migration, an insect that selects its host plant, an animal that avoids a predator, a male that displays to a female – they all act purposefully because they have been programmed to do so. When I speak of the programmed 'individual' I do so in a broad sense. A programmed computer itself is an 'individual' in this sense, but so is, during reproduction, a pair of birds whose instinctive and learned actions and interactions obey, so to speak, a single programme.

The completely individualistic and yet also species-specific DNA code of every zygote (fertilized egg cell), which controls the development of the central and peripheral nervous systems, of the sense organs, of the hormones, of physiology and morphology, is the *programme* for the behaviour computer of this individual.

Natural selection does its best to favour the production of codes guaranteeing behaviour that increases fitness. A behaviour programme that guarantees instantaneous correct reaction to a potential food source, to a potential enemy, or to a potential mate will certainly give greater fitness in the Darwinian sense than a programme that lacks these properties. Again, a behaviour programme that allows for appropriate learning and the improvement of behaviour reactions by various types of feedbacks gives greater likelihood of survival than a programme that lacks these properties.

The purposive action of an individual, in so far as it is based on the properties of its genetic code, therefore is no more nor less purposive than the actions of a computer that has been programmed to respond appropriately to various inputs. It is, if I may say so, a purely mechanistic purposiveness.

We biologists have long felt that it is ambiguous to designate such programmed, goal-directed behaviour 'teleological', because the word *teleological* has also been used in a very different sense for the final stage in evolutionary adaptive processes. When Aristotle spoke of final causes he was particularly concerned with the marvellous adaptations found throughout the plant and animal kingdom. He was concerned with what later authors have called design or plan in nature. He ascribed to final causes not only mimicry or symbiosis but all the other adaptations of animals and plants to each other and to their physical environment. The Aristotelians and their successors asked themselves what goal-directed process could have produced such a well-ordered design in nature.

It is now evident that the terms *teleology* and *teleological* have been applied to two entirely different sets of phenomena. On one hand is the production and perfecting throughout the history of the animal and plant kingdoms of ever-new programmes and of ever-improved DNA codes of information. On the other hand, there is the testing of these programmes and the decoding of these codes throughout the lifetime of each individual. There is a fundamental difference between, on the one hand, end-directed behavioural activities or developmental processes of an individual or system, which are controlled by a programme, and, on the other hand, the steady improvement of genetic codes. This genetic improvement is evolutionary adaptation controlled by natural selection.

In order to avoid confusion between the two entirely different types of end direction, Pittendrigh [6] has introduced the term *teleonomic* as a descriptive term for all end-directed systems 'not committed to Aristotelian teleology'. Not only does this negative definition place the entire burden on the word *system*, but it makes no clear distinction between the two teleologies of Aristotle. It would seem useful to restrict the term *teleonomic* rigidly to systems operating on the basis of a programme, a code of information. Teleonomy in biology designates 'the apparent purposefulness of organisms and their characteristics', as Julian Huxley expressed it [7].

Such a clear-cut separation of teleonomy, which has an analysable physico-chemical basis, from teleology, which deals more broadly with the overall harmony of the organic world is most useful because these two entirely different phenomena have so often been confused with each other.

The development or behaviour of an individual is purposive, natural selection is definitely not. When MacLeod [8] stated 'What is most challenging about Darwin, however, is his reintroduction of purpose into the natural world', he chose the wrong word. The word *purpose* is singularly inapplicable to evolutionary change, which is, after all, what Darwin was considering. If an organism is well adapted, if it shows superior fitness, this is not due to any purpose of its ancestors or of an outside agency, such as 'Nature' or 'God', who created a superior design or plan. Darwin 'has swept out such finalistic teleology by the front door', as Simpson [9] has rightly said.

We can summarize this discussion by stating that there is no conflict between causality and teleonomy, but that scientific biology has not found any evidence that would support teleology in the sense of various vitalistic or finalistic theories [9, 10]. All the so-called teleological systems which Nagel discusses [11] are actually illustrations of teleonomy.

▸ *The Problem of Prediction.* The third great problem of causality in biology is that of prediction. In the classical theory of causality the touchstone of the goodness of a causal explanation was its predictive value. This view is still maintained in Bunge's modern classic [12]: 'A theory can predict to the extent to which it can describe and explain.' It is evident that Bunge is a physicist: no biologist would have made such a statement. The theory of natural selection can describe and explain phenomena with considerable precision, but it cannot make reliable predictions, except through such trivial and meaningless circular statements as, for instance, 'the fitter individuals will on the average leave more offspring'. Scriven [13] has emphasized quite correctly that one of the most important contributions to philosophy made by the evolutionary theory is that it has demonstrated the independence of explanation and prediction.

Although prediction is not an inseparable concomitant of causality, every scientist is nevertheless happy if his causal explanations simultaneously have high predictive value. We can distinguish many categories of prediction in biological explanation. Indeed, it is even doubtful how to define 'prediction' in biology. A competent zoogeographer can predict with high accuracy what animals will be found on a previously unexplored mountain range or island. A paleontologist likewise can predict with high probability what kind of fossils can be expected in a newly accessible geological horizon. Is such correct guessing of the results of past events genuine prediction? A similar doubt pertains to taxonomic predictions. as discussed in the next paragraph. The term *prediction* is, however, surely legitimately used for future events. Let me give you four examples to illustrate the range of predictability:

1. *Prediction in classification.* If I have identified a fruit fly as an individual of *Drosophila melanogaster* on the basis of bristle pattern and the proportions of face and eye, I can 'predict' numerous structural and behavioural characteristics which I will find if I study other aspects of this individual. If I find a new species with the diagnostic key characters of the genus *Drosophila*, I can at once 'predict' a whole set of biological properties.

2. *Prediction of most physicochemical phenomena on the molecular level.* Predictions of very high accuracy can be made with respect to most biochemical unit processes in organisms, such as metabolic pathways, and with respect to

biophysical phenomena in simple systems, such as the action of light, heat, and electricity in physiology.

In examples 1 and 2 the predictive value of causal statements is usually very high. Yet there are numerous other generalizations or causal statements in biology that have low predictive values. The following examples are of this kind.

3. *Prediction of the outcome of complex ecological interactions.* The statement 'An abandoned pasture in southern New England will be replaced by a stand of grey birch (*Betula populifolia*) and white pine (*Pinus strobus*)' is often correct. Even more often, however, the replacement may be an almost solid stand of *P. strobus*, or *P. strobus* may be missing altogether and in its stead will be cherry (*Prunus*), red cedar (*Juniperus virginianus*), maples, sumac and several other species.

Another example also illustrates this unpredictability. When two species of flour beetles (*Tribolium confusum* and *T. castaneum*) are brought together in a uniform environment (sifted wheat flour), one of the two species will always displace the other. At high temperatures and humidities *T. castaneum* will win out; at low temperatures and humidities *T. confusum* will be the victor. Under intermediate conditions the outcome is indeterminate and hence unpredictable (Table 1) [14].

TABLE 1

Two species of *Tribolium* in competition (from Park [14])

Condition		*Replicas (No.)*	*Victorious species (No. of trials)*	
Temp. (°C)	*Humidity (%)*		T. confusum	T. castaneum
34	70	30	—	30
29	70	66	11	55
24	70	30	21	9
34, 29	30	60	53	7
24	30	20	20	—

4. *Prediction of evolutionary events.* Probably nothing in biology is less predictable than the future course of evolution. Looking at the Permian reptiles, who would have predicted that most of the more flourishing groups would become extinct (many rather rapidly), and that one of the most undistinguished branches would give rise to the mammals? Which student of the Cambrian fauna would have predicted the revolutionary changes in the marine life of the subsequent geological

eras? Unpredictability also characterizes small-scale evolution. Breeders and students of natural selection have discovered again and again that independent parallel lines exposed to the same selection pressure will respond at different rates and with different correlated effects, none of them predictable.

As is true in many other branches of science, the validity of predictions for biological phenomena (except for a few chemical or physical unit processes) is nearly always statistical. We can predict with high accuracy that slightly more than 500 of the next 1000 newborns will be boys. We cannot predict the sex of a particular unborn child.

▶ *Reasons for Indeterminacy in Biology.* Without claiming to exhaust all the possible reasons for indeterminacy, I can list four classes. Although they somewhat overlap each other, each deserves to be treated separately.

1. *Randomness of an event with respect to the significance of the event.* Spontaneous mutation, caused by an 'error' in DNA replication, illustrates this cause for indeterminacy very well. The occurrence of a given mutation is in no way related to the evolutionary needs of the particular organism or of the population to which it belongs. The precise results of a given selection pressure are unpredictable because mutation, recombination, and developmental homeostasis are making indeterminate contributions to the response to this pressure. All the steps in the determination of the genetic contents of a zygote contain a large component of this type of randomness. What we have described for mutation is also true for crossing over, chromosomal segregation, gametic selection, mate selection, and early survival of the zygotes. Neither underlying molecular phenomena nor the mechanical motions responsible for this randomness are related to their biological effects.

2. *Uniqueness of all entities at the higher levels of biological integration.* In the uniqueness of biological entities and phenomena lies one of the major differences between biology and the physical sciences. Physicists and chemists often have genuine difficulty in understanding the biologist's stress of the unique, although such an understanding has been greatly facilitated by the developments in modern physics. If a physicist says 'ice floats on water', his statement is true for any piece of ice and any body of water. The members of a class usually lack the individuality that is so characteristic of the organic world, where all individuals are unique; all stages in the life cycle are unique; all populations are unique; all species and higher categories are unique; all interindividual contacts are unique; all natural associations of species are unique; and all evolutionary events are unique. Where these statements are applicable to man, their validity is self-evident. However, they

are equally valid for all sexually reproducing animals and plants. Uniqueness, of course, does not entirely preclude prediction. We can make many valid statements about the attributes and behaviour of man, and the same is true for other organisms. But most of these statements (except for those pertaining to taxonomy) have purely statistical validity. Uniqueness is particularly characteristic for evolutionary biology. It is quite impossible to have for unique phenomena general laws like those that exist in classical mechanics.

3. *Extreme complexity.* The physicist Elsässer stated in a recent symposium: '[an] outstanding feature of all organisms is their well-nigh unlimited structural and dynamical complexity.' This is true. Every organic system is so rich in feedbacks, homeostatic devices, and potential multiple pathways that a complete description is quite impossible. Furthermore, the analysis of such a system would require its destruction and would thus be futile.

4. *Emergence of new qualities at higher levels of integration.* It would lead too far to discuss in this context the thorny problem of 'emergence'. All I can do here is to state its principle dogmatically: 'When two entities are combined at a higher level of integration, not all the properties of the new entity are necessarily a logical or predictable consequence of the properties of the components.' This difficulty is by no means confined to biology, but it is certainly one of the major sources of indeterminacy in biology. Let us remember that indeterminacy does not mean lack of cause, but merely unpredictability.

All four causes of indeterminacy, individally and combined, reduce the precision of prediction.

One may raise the question at this point whether predictability in classical mechanics and unpredictability in biology are due to a difference of degree or of kind. There is much to suggest that the difference is, in considerable part, merely a matter of degree. Classical mechanics is, so to speak, at one end of a continuous spectrum, and biology is at the other. Let us take the classical example of the gas laws. Essentially they are only statistically true, but the population of molecules in a gas obeying the gas laws is so enormous that the actions of individual molecules become integrated into a predictable – one might say 'absolute' – result. Samples of five or 20 molecules would show definite individuality. The difference in the size of the studied 'populations' certainly contribute to the difference between the physical sciences and biology.

▸ *Conclusions.* Let us now return to our initial question and try to summarize some of our conclusions on the nature of the cause-and-effect relations in biology.

1. Causality in biology is a far cry from causality in classical mechanics.

2. Explanations of all but the simplest biological phenomena usually consist of sets of causes. This is particularly true for those biological phenomena that can be understood only if their evolutionary history is also considered. Each set is like a pair of brackets which contains much that is unanalyzed and much that can presumably never be analysed completely.
3. In view of the high number of multiple pathways possible for most biological processes (except for the purely physicochemical ones) and in view of the randomness of many of the biological processes, particularly on the molecular level (as well as for other reasons), causality in biological systems is not predictive, or at best is only statistically predictive.
4. The existence of complex codes of information in the DNA of the germ plasm permits teleonomic purposiveness. On the other hand, evolutionary research has found no evidence whatsoever for a 'goal-seeking' of evolutionary lines, as postulated in that kind of teleology which sees 'plan and design' in nature. The harmony of the living universe, so far as it exists, is an a posteriori product of natural selection.

Finally, causality in biology is not in real conflict with the causality of classical mechanics. As modern physics has also demonstrated, the causality of classical mechanics is only a very simple, special case of causality. Predictability, for instance, is not a necessary component of causality. The complexities of biological causality do not justify embracing non-scientific ideologies, such as vitalism or finalism, but should encourage all those who have been trying to give a broader basis to the concept of causality.

References

1. E. Nagel. Lecture presented at the Massachusetts Institute of Technology in the 1960–61 Hayden Lectures series.
2. M. Delbrück. *Trans. Conn. Acad. Arts Sci., 33*, 173 (1949).
3. J. Loeb. *The Organism as a Whole* (Putnam: New York 1916).
4. M. Scriven. Unpublished manuscript.
5. C. Bernard. *Leçons sur les phénomènes de la vie* (1885), vol. 1.
6. C. S. Pittendrigh, in *Behavior and Evolution*, A. Roe and G. G. Simpson, Eds. (Yale Univ. Press: New Haven, Conn., 1958), p. 394.
7. J. Huxley. *Zool. Jahrb. Abt. Anat. u. Ontog. Tiere, 88*, 9 (1960).
8. R. B. MacLeod. *Science, 125*, 477 (1957).
9. G. G. Simpson. *Ibid., 131*, 966 (1960).
10. G. G. Simpson. *Sci. Monthly, 71*, 262 (1950). L. F. Koch. *Ibid., 85*, 245 (1957).
11. E. Nagel. *The Structure of Science* (Harcourt Brace and World Inc.: New York 1961).
12. M. Bunge. *Causalty* (Harvard Univ. Press: Cambridge, Mass., 1959), p. 307.
13. M. Scriven. *Science, 130*, 477 (1959).
14. T. Park. *Physiol. Zool., 27*, 177 (1954).

Comments by C. H. Waddington

Reprinted from Science, 135, *976 (1962)*

The very interesting article by Mayr [1] seems to call for some comment. Mayr distinguishes between an unacceptable and an acceptable form of finalistic explanation; the former is teleological or vitalistic, the latter 'teleonomic', or, in the phrase I have used, 'quasi-finalistic'. But having made this distinction between types of hypothesis, Mayr proceeds to assert that there is a corresponding cleavage between types of phenomena. 'The development or behaviour of an individual', he writes, 'is purposive' (that is, in the acceptable sense), 'natural selection is definitely not'; and by 'is not' he seems to mean 'cannot be', since in another place he writes, 'Historical processes, however, can *not* act purposefully.' I have for some years been urging that quasi-finalistic types of explanation are called for in the theory of evolution as well as in that of development [2].

If any process is set going (for example, if two chemical substances are allowed to start reacting with one another) it will eventually reach some end. The question of 'finalism' arises when there is something interesting about the end – in particular, when it is both complex and definite in character. We then have three main types of explanation available: (i) that the end itself acts as a cause, directing the process so that it terminates at the predetermined end state; this is Aristotelian finalism, which we reject because it involves a concept of causation quite outside our accepted range of ideas; (ii) that some non-material agency directs the process to the predetermined end; this is 'vitalism', which we also reject; (iii) that the end state of the process is determined by its properties at the beginning; this is 'mechanism', and our recent experience of such mechanical systems as computers has led us to realize that it is a more powerful type of hypothesis than it had previously appeared to be. We can set up a process in such a way that it will reach an assigned end state by building into the initial situation a set of conditions which act as a 'programme', and by providing suitable negative feedback relations to bring the process back on to the right course if it should diverge from it. Conversely if we find any process to be characterized by a programme and feedbacks, we can deduce that it will proceed towards some end which should in principle be ascertainable from the nature of the programme and the feedbacks (the degree of precision with which the end is determined will depend, of course, on the particular characteristics of the programme and feedbacks).

Mayr accepts the theory [3] that ontogenetic development depends on a quasi-finalistic mechanism of this type, the programme and feedback relations both being incorporated in the genotype which has been moulded by natural selection. But there is nothing in the nature of such quasi-finalistic mechanisms which makes it impossible to suppose that the evolutionary process itself is also of this kind. It is obviously characterized by a programme, that involved in the theorem of natural selection. This in itself suffices to determine, to a certain degree, the nature of the end towards which evolution will proceed; it must result in an increase in the efficiency of the biosystem as a whole in finding ways of reproducing itself. And there are, surely, many feedback relations which will serve to determine ends in a more precise fashion. The two to which I have previously directed attention are (i) that involved in the fact that an organism's behaviour influences the nature of selection pressures which will operate on it (loosely, an animal selects its environment before its environment selects it), and (ii) that arising from the fact that selection of previous generations for stability or flexibility of development will influence the type of phenotypic effect likely to be produced by new mutation. There are certainly many others. For instance, increasing phenotypic diversification of a population to fit it to deal with different habitats will eventually be counter-balanced by the development of barriers to interbreeding of the different varieties.

It seems to me that it is becoming inadequate to point out, as Mayr does, that natural selection is not purposive. In itself it is of course no more purposive than is the process of formation of interatomic chemical bonds. But just as the latter process is the basic mechanism underlying the protein syntheses which are integrated into the quasi-finalistic mechanism of embryonic development, so natural selection is the basic mechanism of another type of quasi-finalistic mechanism, that of evolution. The need at the present time is to use our newly won insights into the nature of quasi-finalistic mechanisms to deepen our understanding of evolutionary processes.

C. H. Waddington
Institute of Animal Genetics
University of Edinburgh, Scotland

References

1. E. Mayr. *Science, 134,* 1501 (1961).
2. C. H. Waddington. *The Strategy of the Genes* (Allen and Unwin: London 1957); *The Nature of Life* (Allen and Unwin: London; Atheneum: New York 1961).
3. C. H. Waddington. *Organisers and Genes* (Cambridge Univ. Press: Cambridge 1940).

An Approach to a Blueprint for a Primitive Organism

A. G. Cairns-Smith
University of Glasgow

Haldane [1], in attempting to draw up a blueprint for the first organism, suggested that the simplest possible arrangement might have consisted of an RNA gene specifying a single enzyme. He considered that even this would have been too improbable a structure to turn up without the operation of some hitherto unnoticed principle which makes biological systems intrinsically more probable than they would otherwise seem to be. Von Neumann's discussion [2] of 'self-reproduction' is similarly discouraging in that it seems to place a high lower limit of necessary complexity on any 'self-reproducing' organization [3].

I will take it that the primitive organisms which interest us conformed both to Muller's general definition of life (Waddington, p. 3) and to Waddington's 'most important fact' that they could take part in the long-term processes of evolution. Haldane's discussion I will take as an indication that the very first organisms were not constructed along the lines of modern organisms: as discussed by Pattee, there is no logical reason for insisting that the same molecules that are universal in modern organisms were necessarily present, or similarly functional, in the most primitive forms. The well-known syntheses of 'biochemicals' under assumed primitive earth conditions do not present any clear argument for biochemical similarity between modern and primitive life-forms: organisms evolving in an environment containing aminoacids, purines, etc., would have been likely to incorporate these molecules in their structures at some stage, but not necessarily at the point of origin.

The difficulties suggested by von Neumann's analysis will be taken as an indication that the replication of primitive organisms was largely or wholly imposed on them by their environment – that the process of replication *per se* did not require the pre-existence of any elaborate internal instructions or machinery within the organisms. We might think of the replication of the unit cell of a crystal, or better the replication of a pattern of dislocations, during the growth of a crystal. The very commonness of crystal growth processes, which can be maintained indefinitely within a suitable maintained flow environment, is an indication that no elaborate specific instructions are required for their operation. It may be, as Waddington says (p. 4), that a simple dislocation is not interesting enough to be called an organism; but I think we must accept that life – 'interesting' life – arose from systems that were 'uninteresting' int he sense that any functional specificity which they contained must have been very limited indeed. Rather than consider theoretical models of replication processes that closely mirror those of modern organisms we should perhaps look very hard at the simple processes of replication which already exist in profusion in the physico-chemical world, and to consider these not simply as models, but as potential ancestors. Our problem may be wholly one of evolution rather than of origin: it may, for example, be that of finding an evolutionary route from some already well-known crystal to the kind of organisms that presently inhabit the earth.

▸ *A gene in a stream?* A maintained environmental flow of some sort is, of course, essential for life. Modern organisms are more or less directly driven by the flow of photons from the sun. Sherrington [4] compared organisms to eddies in a stream. We might elaborate this analogy and say that the eddies themselves are phenotypes, produced by genotypes consisting of the more permanent stones and banks of the stream which control the form of the eddies. For the analogy to be more complete we would need something like a replicable stone, and a situation in which the specific pattern of eddies, which a stone of a specific shape produced, could affect the survival potential of the stone in the stream.

Considered at a molecular level, this analogy can be made more realistic and provide us with a possible model for a primitive organism. The stream we take as literal, consisting of a maintained hydrodynamic flow on or near the surface of the primitive earth. The 'stone' becomes a crystallite containing a pattern of imperfections. The stream is (sometimes) supersaturated with respect to the components of the crystallite so that new crystallites are formed, through crystal growth and cleavage, with similar imperfection patterns. The 'eddies' result from pattern-dependent interactions between the crystallites and other dissolved components in the stream (e.g. they consist of a changing population of more or less specifically adsorbed molecules which tend to anchor the crystallite in the part of the stream where crystal growth is favoured, or which protect the crystallite against dissolution when the stream intermittently ceases to be saturated, or which accelerate the processes of crystal growth and cleavage).

Elsewhere [5] it has been argued that clay crystallites may have constituted the genes of primitive organisms, with survival-promoting instructions written in the form of patterns of internal cation substitutions, and with phenotypes consisting of adsorbed, abiogenically formed, organic molecules. Whether this is a practical possibility will depend, among other things, on whether internal patterns of cation substitutions in clays tend to replicate during clay synthesis from solution. We will pursue here the more general idea that our ultimate ancestor was 'a gene in a stream' without supposing that the gene in question was necessarily a clay or, for that matter, a crystallite of any kind.

▸ *A general specification for an evolving physico-chemical system.* We require a system that can go indefinitely round the Darwinian cycle illustrated in figure 1. It

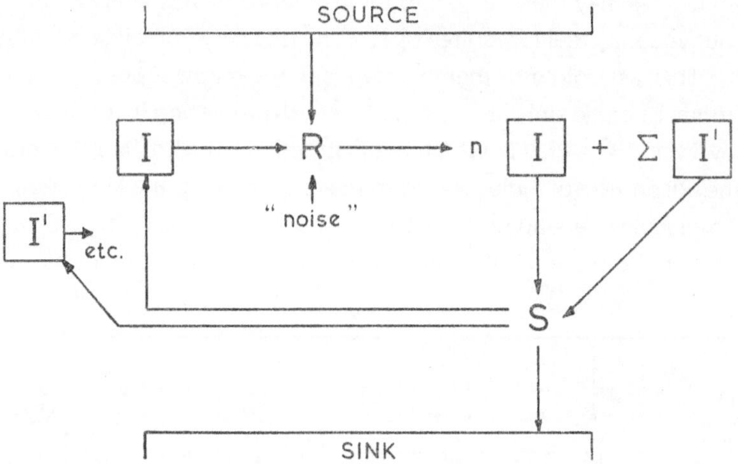

FIGURE 1
Darwinian cycle for 'type 2' organism (see text). Occasional errors ('noise') in the replication gives rise to modified patterns, some of which, e.g. I', may start a competing cycle

will consist of a population of microscopic boxes holding various patterns, e.g. I, and existing in a flow environment which operates on the boxes in two ways: (fi) to replicate the patterns (usually but not quite always accurately) through the synthesis of more boxes, and (ii) to select the boxes according to the patterns which they contain. Some patterns may be selected because they accelerate the replication process itself. Where this is the only kind of selection, as in the preferential formation of growth-promoting dislocations in a crystal, we might say that we have a 'type 1' organism. In a 'type 2' organism (figure 1) there exists alternatively, or in addition, a selection stage which is independent of the replication process, i.e. the pattern may specifically improve the chances that the box holding it will survive. We might imagine here a fluctuating environment which favours at different times (i) the synthesis of boxes (replicating conditions, R), and (ii) destruction of boxes (selecting conditions, S). The evolution of the system would then tend to generate patterns which (i) favoured synthesis and (ii) resisted destruction of boxes, i.e. which would be carried most efficiently round the cycle. These patterns would represent genetic instructions related to the particular environment.

We see the environment, then, as the active part of the composite system, generating instructions written in the form of a series of stable patterns at the molecular level. Thus we reverse the more usual view and consider the environment ('the stream') as essentially dynamic, and the organism ('the gene') as essentially static. We may then be able to think in terms of very simple initial organisms, but you may feel it requires us to think of a very complicated initial environment! This last point is rather intuitive, but one might suppose that a simple environment could only elicit simple survival-promoting instructions. Suppose, however, that a way in which the instructions in a box are self-protecting is through the adsorption of molecules from the environment, the kind and pattern of adsorbed molecules depending on the instructions in the box (a 'type 3' organism, see figure 2), then, from the point of view of the box, the environment has now

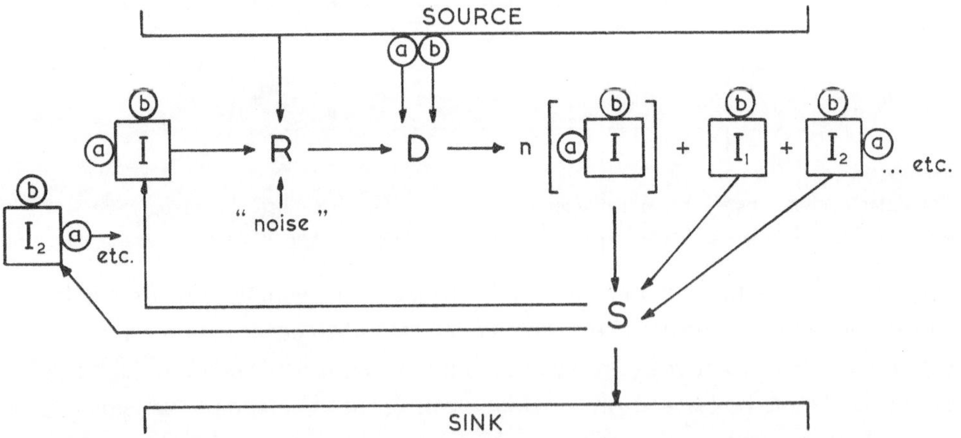

FIGURE 2
Cycle for an 'adsorption organism' including a development stage, D, consisting of instruction-dependent adsorption from the environment of protective patterns of molecules, e.g. a and b

changed, giving rise to the possible evolution of new instructions that are relevant within this changed environment. These new instructions would again change the immediate environment of the boxes, creating yet further possibilities, and so on.

The initial flow environment, then, which generates the initial instructions need not be very complex provided that, at least partly, these instructions react back on the environment and change it – particularly if they have the effect of producing a persistent modification of the environment in their own vicinity, i.e. a phenotype. (It rather looks as if a phenotype will emerge, whether you want it or not, even from the most strictly genetical view, if you insist that any system to be called an organism must be capable of indefinite evolution, i.e. if you add Waddington's 'most important fact' to Muller's definition.)

▸ *Is 'genetic metamorphosis' possible?* The boxes in figures 1 and 2 are made of some kind of 'genetic material'. There is a general consideration which suggests that this material was, for primitive organisms, neither DNA nor anything like it: the features that would be required in a primary genetic material, which must operate 'in the open' with practically no ancillary machinery, are likely to be very different from those which would be most effective in a highly evolved organism in which replication of the genes and development of phenotype can occur with the assistance of elaborate pre-formed machinery – as in the cytoplasm of modern cells. The specification for 'the best possible' genetic material would change radically during the early stages of evolution. Whether the actual material would change would depend on whether a suitable changeover mechanism existed. It has been suggested in discussion at this symposium that such a process may be logically impossible. If this is so, then nucleic acid, as a single 'compromise' material, must have operated throughout. I would like to indicate now that a *'genetic metamorphosis'* is not only possible, but that early organisms may well have been constructed along lines that provided a ready mechanism for it to have taken place.

Relatively minor changes in a genetic material, that we will call *'genetic modification'*, may not present any very fundamental difficulty. Haldane [1] suggested, for example, that RNA preceeded DNA as the genetic material: this is not too difficult to envisage, since RNA and DNA are 'compatible'. But such a change is simply to alter the colour of a blueprint ink. We would like to know whether evolution could have proceeded through a change from blueprints to magnetic tape! For a radical change in the method of coding, a sequence of genetic modifications through pairs of mutually compatible systems becomes a somewhat implausible mechanism. It would seem, however, that there exists in theory an alternative mechanism which does not require chemical similarity between the different genetic materials, because it does not involve any transcription of instructions between them.

▸ *'Heterogenetic' organisms.* Waddington (p. 16) and Sager have considered the possibility that in modern organisms there may be more than one process for the transmission of biological information. It is interesting to consider the possibility that primitive organisms evolved through strongly *'heterogenetic'* forms, i.e. forms containing more than one kind of genetic material.

First we may ask why modern organisms are, at least mainly, *'homogenetic'*.

There may be a formal simplicity in homogenetic systems, but in modern organisms this seems to be a rather sophisticated kind of 'simplicity' which depends on the remarkable versatility of the single system; a versatility which in its turn seems to depend on a division of function between two quite different kinds of molecule, DNA being particularly good at storage and transmission of genetic instructions and protein particularly effective in translating these instructions into a wide range of structures and activities. The modern 'simplicity' depends on the existence of elaborate translating machinery. (It might be formally simpler if all the information that we use in everyday life were stored in a single form – say on magnetic tape – but until everyone has suitable equipment it will still be more convenient to use books, gramophone records, films, etc.)

Pirie [6] has argued that the modern biochemical uniformity – with all organisms using essentially the same system – was the result of a long process of selective evolution from numerous chemically different alternative systems. We will consider here implications of biochemical diversity not so much *between* primitive organisms as *within* them.

▸ *A mechanism for genetic metamorphosis*. The initial organisms may well have been homogenetic with a very simple direct and limited survival technique (cf. [5]). There are two processes that could increase the versatility of a primitive genetic system: firstly *'genetic extension'*, involving the evolution of increasingly indirect modes of action, e.g. (cf. figure 2) the genes not only adsorb molecules from their environment to protect them against destruction, but they form catalytic centres on their surfaces which allow new and more effective molecules to be formed. Further extension could arise if some of the new molecules were themselves control devices, e.g. catalysts. Genetic extension could increase the versatility of a primitive genetic material, but there is no particular reason to suppose that such a material – one that is a good starter – would be particularly good at evolving very far in this way. A second way of increasing versatility would be for the organisms to become heterogenetic. One might guess that before the appearance of any single highly versatile system, early organisms, although perhaps not the very first, used a number of different, and more or less directly acting, genetic materials each able to form different kinds of biologically useful structures. Further, the more chemically *different* these materials were, the greater would be the range of functions available and the less would be the chance of confusion between them. We might suppose, for example, that one substance could both replicate and act as a membrane or gel-forming fibre, another replicate and act as a catalyst for one kind of reaction, yet another for another kind of reaction, and so on. Each member might be thought of as an organism (or sub-organism) living symbiotically with the others. One might picture the total organism as a jelly of indefinite dimensions containing large numbers of each of the different members. Reproduction of the total organism could occur at first through more or less haphazard breaking off of pieces of the jelly that happened to contain a representative selection of the members. More systematic reproduction involving synchronization of the replication of the genetic members and their spatial organization into cells might appear later but would not be an initial requirement. (A group of organisms may live together to their mutual advantage without closely synchronizing their replication or having any very elaborate spatial organization: in so far as the members of the group are interdependent a rough balance of numbers, within a given zone, will tend to be maintained.)

In the early stages, factors tending to increased heterogenetic character might be expected to predominate with a relatively high probability that a different genetic material, with a different range of immediate functions, would have something to contribute to an organism working a non-versatile system. Later, reverse factors tending back towards homogenetic character would predominate. The primary one here would be processes of genetic extension in the different systems leading to overlap between the kinds of functions which they could perform, and a resulting competition between them. Whereas the primitive genetic material would be selected for its directness of action (no need for preformed machinery), advantage in the later competition would tend to arise from indirectness of action in so far as this could lead to increased versatility.

We may summarize the total situation envisaged as follows. Our ultimate ancestor was homogenetic; its genetic material was simple, rugged, and direct in its limited effects: this evolved eventually into modern organisms which are also (mainly) homogenetic, but whose genetic material is complex, delicate, and indirect in its versatile effects. The 'metamorphosis' occurred through an extended evolution during which the original organisms became heterogenetic. These different genetic materials within the same organism evolved in a competition in which the modern rather than the primitive specification was increasingly favoured. The primitive genetic material was thus one of the casualties, and the eventual 'winner' bore little if any chemical resemblance to it.

▸ *Primitive organisms as 'starters'*. Figure 3 incorporates a primitive cycle operating with a genetic material X, i.e. R_XS_X. This cycle is a simplified version of the kinds shown in figures 1 and 2. We may consider the general question of how the evolution of instructions in X could give rise to, or assist, the evolution of instructions in some other genetic material, say Y, without direct transcription between the two materials. Genetic metamorphosis, as discussed above, is the most extreme of three possible related processes by which $R_X S_X$ might act as a starter for $R_Y S_Y$. One should not think of these as mutually exclusive.

(i) *'Saprophitic'*. As the evolution of primitive forms proceeded, the range of substances available on the earth would change. This in itself would increase the probability of alternative genetic materials arising at some stage. Moreover the substances that became available would probably include resolved optically active materials more suitable for the formation of genetic polymers than the racaemic mixtures that would have been usual in the original environment. $R_X S_X$ might thus give rise to a quite separate cycle, $R_Y S_Y$, by providing for the first time a suitable 'food'.

(ii) *'Parasitic'*. If a new genetic material arose within the phenotype of a primary form, this might provide a particularly easy environment for the establishment of a critical cycle (i.e. one that produces on average enough viable offspring to continue indefinitely) – not only because the new life-form could make use of a novel stockroom of molecules (and perhaps higher order structures such as tubes and grooved surfaces), but also because aspects of the activity of the primary form might assist in the establishment of a critical cycle. For example, the primary form might have developed the characteristics of a 'Goodwin oscillator' (cf. Goodwin, p. 137) as a means of becoming less dependent on environmental fluctuations to control its genetic replication, and thus provide a controlled fluctuating environment for the parasite. Such a parasitic association might be permanent ($R_{Y(X)} S_{Y(X)}$), or assist only part of the life cycle of the parasite ($R_Y S_{Y(X)}$ or $R_{Y(X)} S_Y$). We should reckon, then, with the possibility that our ultimate ancestor was a 'spontaneously generating obligate parasite' which gradually evolved towards independence.

(iii) *'Symbiotic'*. A true genetic metamorphosis is envisaged as proceeding through a situation in which a secondary (obligate) sub-organism adds slightly to the competence of a primary form. The mutual evolution of coupling instructions in the different genetic materials could then be expected – it would be to their mutual advantage to stick together, perhaps quite literally. The composite system would then evolve through selection of mutations in either of its genetic materials – it would have become a 'digenetic' organism. The cycles $R_{Y(X)} S_{Y(X)}$ and $R_{X(Y)} S_{X(Y)}$ (figure 3) could then be regarded as a single cycle. Assuming, for simplicity, that no other material is involved, then a complete metamorphosis would result if Y turns out to have a greater evolutionary potential than X (by being capable of greater genetic extension) to such an extent that it can eventually dispense with X. An $R_Y S_Y$ cycle would then be established, together with the relatively high level of instructions and preformed phenotypic machinery the lack of which had made Y incompetent as a primitive genetic material.

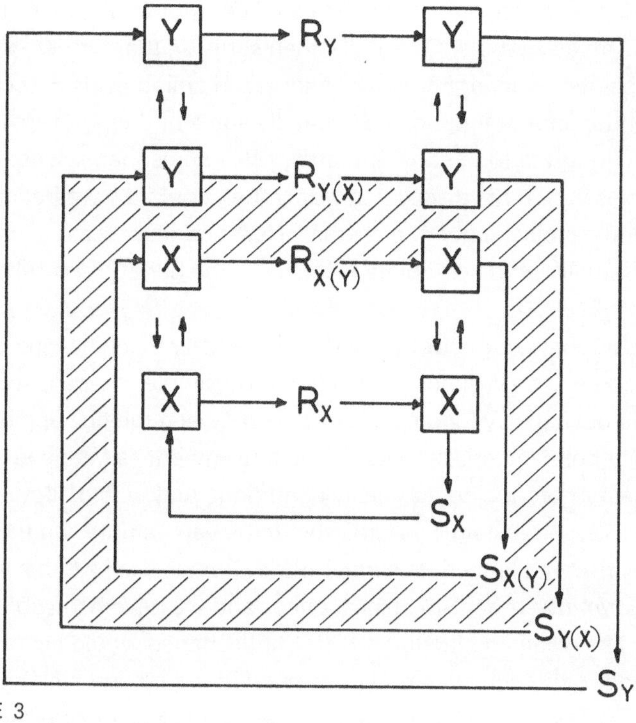

FIGURE 3

Diagram illustrating ways in which a 'self-starting' cycle (highly simplified) operating with a genetic material X could assist the establishment of a critical cycle by some other material Y (see text). '$R_{Y(X)}$' and '$S_{Y(X)}$' mean respectively 'replication and selection of genetic material Y, with the assistance of X and/or its associated phenotype'. Similarly for $R_{X(Y)}$ and $S_{X(Y)}$. Small vertical arrows indicate possibility of partial associations. Hatching represents coupled cycle of a 'digenetic' organism

▸ *What about nucleic acid?* The appearance of nucleic acid, or of a precursor capable of evolving into nucleic acid simply by genetic modification, was certainly a crucial stage in evolution. Some may like to call it 'the origin of life', although (cf. Pirie, 1959) this is surely a rather prejudiced point of view. According to the view which has been pursued here, this 'origin of life' would have been a very gradual process. It may have started with something like RNA being formed within the phenotype of an evolved primitive form and operating as a subsidiary genetic material in the performance of some quite minor function – this function (whatever it was) depending perhaps on the ability to form a specific sequence-dependent tertiary structure through intra-molecular base-pairing. Genetic extension might then have occurred through specific associations between such RNA organelles and small molecules in their surroundings – particularly with amino-acids – to give objects with some more sophisticated contribution to make to the heterogenetic organism as a whole. One might then suppose that further extension occurred through specific coupling together of adsorbed amino acids to give polypeptides with independent functions. Eventually some of the polypeptides would become big enough to form coherent tertiary structures themselves – and subsequently enzymes would be 'discovered'. Only then could it have become apparent (to a primitive molecular biologist) that RNA had discovered a 'winning' system, capable eventually of the complete takeover of the genetic control of the whole organism.

Whatever the evolutionary route for the modern hereditary machine, and the above 'genetic ribosome' idea is almost certainly far too simple, it would seem that there must have been a very long period of time between the first appearance of nucleic acid as *a* genetic material and its final establishment as *the* genetic material. The prospect of producing artificially conditions that would allow such a process to occur in a reasonable time would seem to be very remote. On the other hand, that creative, but limited, character Blind Chance seems to have managed to form a *primitive* organism at least once; and we have been able to put together things like insulin molecules which Blind Chance would almost certainly have found too difficult even under the most idealized conditions on the primitive earth (cf. [7, 8]). A primitive organism, then, should be quite easy to make – if we know what to do. First we should try to think of any practical physico-chemical system capable of indefinite Darwinian evolution. And we should not insist that our blueprint must include nucleic acid – or, for that matter, any organic polymer at all.

References

1. J. B. S. Haldane in (S.W. Fox, ed.) *The Origins of Prebiological Systems* (New York: Academic Press 1965).
2. J. von Neumann (1948) in (A. H. Taub, ed.) *John von Neumann Collected Works* vol 5 (Oxford: Pergamon Press 1963) p. 288
3. H. H. Pattee *Advance. Enzymol. 27* (1965) 381.
4. C. Sherrington *Man on his Nature* (Cambridge: University Press 1940) ch. 3.
5. A. G. Cairns-Smith *J. Theoret. Biol. 10* (1966) 53.
6. N. W. Pirie in (F. Clark and R. L. M Synge, eds. *The Origion of Life on the Earth* (New York: Pergamon Press 1959) p. 76.
7. M. Dixon and E. C. Webb *Enzymes* (New York: Academic Press 1958) p. 666.
8. H. H. Pattee *Biophys. J. 1* (1961) 683.

The Physical Basis of Coding and Reliability in Biological Evolution

H. H. Pattee
Stanford University

The following correspondence is relevant to Dr Pattee's paper, printed below:

17th February 1967

Dr Howard Pattee
Center for Theoretical Studies
University of Miami

Dear Howard,

I hope that in the MS you are preparing for us you will not mind throwing in a few quite elementary paragraphs for the benefit of the inadequately educated biologist. For instance, I am still not a bit clear about the meaning of 'a non-holonomic constraint'. I think also that you will have to make a more definite bridge between the quantum versus classical thinking in physics and the simple-minded empirical biological approach. For instance, you argue that the simplest classical machines which exhibit an abstractly defined heredity actually perform more like catalysts than like templates; from which you go on to attribute the specificity of base-pairing in DNA replication to the polymerases – but the biologist will be thinking that a gene mutation which is heritable is not, primarily, a change in a polymerase but in a base sequence. I want you to go further into the distinction between 'enzymes as realizers', i.e. which make it possible for adenine to get coupled up with thymine, and 'enzymes as determinants of specificity', i.e. deciding whether adenine gets coupled with thymine or with guanine.

Yours sincerely,

C. H. Waddington

February 21, 1967

Professor C. H. Waddington
Institute of Animal Genetics
Edinburgh

Dear Professor Waddington,

To expand very briefly on your comments, I distinguish hereditary storage *from hereditary* transmission *processes. Storage is entirely time-independent, just like unread books in the stacks or a magnetic tape on the shelf. Mutations can occur during storage, of course, and since DNA is the storage molecule, that is where some mutations can occur. But a mutation in a time-independent storage can have no effect until it is transmitted (replicated, transcribed, or decoded). What I have concluded is that* all hereditary transmission processes must be executed by non-holonomic constraints. *For most biologists, the concept of an allosteric device is almost identical with the idea of non-holonomic constraint, i.e. a many-structured device which changes its structure upon collision with an outside element (activator or substrate), and thereupon causes a specific reaction to speed up. Or, in other words, a certain class of collisions with a non-holonomic or allosteric machine produces a decision and a reaction – 'a controlled action' (not unlike the idea of amplification). Now it is quite likely that DNA has some allosteric or non-holonomic properties (it is a difficult problem to specify when quantum mechanical structures are non-holonomic), but the evidence that I happen to know is more consistent with the enzyme being the overriding decision-making structure in base-pairing during all transmission processes including replication. Certainly the amino acyl synthetases do all the matching of amino acids with the transfer RNAs. I suspect that an enzyme could be designed which matches and polymerizes any two nucleotides in a chain, but of course the resulting storage structure would be relatively unstable compared to the existing pairing. Or, to put it another way, the reliability in cellular DNA replication is, say, one error in* 10^8 *pairings. My guess is that at least 99·9 per cent of this* transmission reliability *resides in the enzyme, not in the hydrogen bond template (which is nevertheless good for storage).*

Many chemists and even biologists regard enzymes primarily as catalysts which now make life go at a pleasant rate, but which might be omitted in principle, especially at the beginning of life, where it is easier to wait on geological time scales for a nucleic acid to replicate or for any reaction to go. I shall argue that this concept is totally false, and that it is only when specific or tactic catalysts execute 'controlled action' in the above sense that hereditary processes (which include all memory-controlled processes) can begin. Furthermore, words like replication, heredity, controlled action, tactic catalysis, and specific enzyme cannot be defined physically without the concept of reliability. (Jack Cowan will agree with me.) It is here that quantum mechanics must be brought into the picture, but that is a longer story with no clear ending at present. The orthodox, mechanical view is probably useful enough up to the problem of the reliability of hereditary storage and transmission. But this is basic.

Sincerely yours,
Howard H. Pattee

THE PHYSICAL BASIS OF CODING AND RELIABILITY

What is a theory of biology? Within the intellectual discipline of the physicist there has developed a belief in the existence of general and universal theories of nature, and it is the search for such theories which may be said to guide and justify the intellectual efforts of the physicist as well as the design of most physics experiments. What a physicist means by a 'good theory' cannot be exhaustively spelled out. Of course it must include 'fitting the data' or 'predicting observations' in some general sense. However, much deeper and more obscure criteria are also applied, often tacitly or intuitively, to evaluate the quality of a physical theory. For example, general theories can never be 'just so' stories which are only built up bit by bit as data accumulate. General physical theories often stem from relatively simple hypotheses which can be checked by experiment, such as the constancy of the speed of light and the discrete energies of photons from atoms, but they must also be founded upon broad principles which express concepts of conservation, invariance, or symmetry. These abstract principles come to be accepted because from our experience we find that in some sense they appear unavoidable. In other words, without such principles it is difficult even to imagine what we mean by a general physical theory of the universe [1].

Traditionally, in biology, the relation of theory to experiment has been more remote. Much of what is sometimes called biological theory appears to the physicist as a 'just so' story, since it is often only a mathematical formalism designed for the practical solution of a specific type of problem and has no direct relation to general physical laws. This situation is often ascribed to basic differences in the subject matter of the physical and life sciences. Perhaps this lack of a basic biological theory is at the root of unresolved historical vitalist-mechanist arguments, since much biological teminology never even makes contact with the language of physics.

But recently, following the so-called molecular biological revolution, there have been many statements that now, at last, the mystery of life has indeed been reduced to physical language and laws. In particular we find biochemists and molecular geneticists claiming that they have shown that normal physical and chemical laws provide a relatively clear and simple basis for understanding heredity and most aspects of metabolism. The Watson-Crick template DNA model is commonly accepted as the central concept which is said to reveal the mystery of heredity [2], and similarly, the detailed structure of proteins has been said to provide a basic understanding of enzyme mechanisms [3]. A common working assumption of molecular biologists is that the remaining problems will be solved by additional experiments. In any case, they do not see any obstacles or essential mysteries on the horizon [4]. This leads to the attitude that biology is explained in terms of ordinary existing physical laws and that therefore no great effort is necessary to apply physical theory to living matter.

On the other hand, in spite of these detailed factual descriptions of polynucleotide and polypeptide interactions in the cell, many physicists as well as biologists remain uneasy. Is this vast amount of phenomenological description really sufficient to support the claim, which is now made even in elementary biology textbooks, that we have a fundamental understanding of living matter in terms of physical laws – that heredity has proven, after all, to be extraordinarily simple, and that the remaining unknowns about living matter are only details to be filled in by more experiment? Can we say with justification that we understand how the laws of physics explain the essential nature of life?

In the remainder of this paper I shall attempt to express why this claim that biology has now been understood in terms of physical laws is not yet convincing. I shall also give some reasons for concluding that the central mysteries of living matter are not to be solved only by collecting more data. Furthermore, I shall propose that even to make a basic distinction between living and nonliving matter some fundamental logical and physical problems remain to be solved at the quantum mechanical level. In particular, I shall argue that any fundamental theory of biology must describe the physical basis of enzymatically controlled hereditary processes which possess the reliability necessary for evolution, and that this will require what amounts to a deeper understanding of the quantum theory of a molecular measurement process.

▸ *Current molecular biological descriptions*. There is no need here to repeat in any detail the descriptions used in modern molecular biology, since so many reviews are now available. By molecular biological description we shall mean the use of such concepts as the template replication of D N A, the transcription of the genetic message from D N A to messenger R N A, the translation of this coded message to amino acids, and the synthesis of proteins [5]. An enormous amount of detail is now known about these processes and much more will undoubtedly be discovered in the near future. The principal question, however, does not have to do with the quantity or quality of these data, but rather with their physical interpretation. In particular we want to discuss whether or not these molecular biological descriptions allow us to conclude that the nature of living matter can now be understood in terms of the general laws of physics.

Normally when the physicist says he understands, say, the chemical bond in terms of general physical laws, he does not mean simply that he is optimistic that chemical bonds are consistent with quantum mechanics or that if he cared to go into the matter he would find no serious problem in describing the chemical bond by the rules of quantum mechanics. On the contrary, although the chemical bond was first recognized and discussed at great length in classical terms, most physicists regarded the nature of the chemical bond as a profound mystery until Heitler and London quantitatively derived the exchange interaction and showed that this quantum mechanical behaviour accounted for the observed properties of valency and stability. On the other hand, it is not uncommon to find molecular biologists using a classical description of D N A replication and coding to justify the statement that living cells obey the laws of physics without ever once putting down a law of physics or showing quantitatively how these laws are obeyed by these processes. Of course, as a speculative prediction such statements are acceptable. But certainly nothing could be less fruitful than allowing this most fundamental and challenging question of whether living matter can be reduced to the basic laws of physics to be obscured by such pronouncements from molecular biologists without some regard for the established language and laws of physics.

▸ *What is the question?* Let us for the moment, however, assume that the experiments of molecular biology and genetics have indeed shown that no detailed process in living matter evades or violates normal laws of physics. If this were the case, does the question of the nature of life appear answered? In other words, even if it were the case that living matter was exactly the same as nonliving matter with respect to description by physical laws, would we then say that we fully understand life in terms of physical laws? No, I think not, because this does not answer the obvious question of why living matter is so conspicuously *different* from nonliving matter. In other words, we do not find the physical similarity of living and nonliving matter so puzzling as the observable differences. Before we can attempt to explain these differences in terms of physical laws we must state clearly what these differences are. Older biology texts usually begin by listing the 'characteristics of life' which may include growth, reproduction, irritability, metabolism, etc., but these are not general enough concepts. What is the most general property of life which distinguishes it from nonliving matter? Certainly the most *general* property is the potential to evolve. Therefore the fundamental question can be restated: Is the process of biological evolution understandable in terms of basic laws of physics?

▸ *Two basic assumptions.* In order to show the difficulties in answering this question let us restate the situation in the form of assumptions:

Assumption A. Living states and nonliving states of matter are in no way distinguishable by their detailed description in terms of initial conditions or elementary laws of motion, i.e. both living and nonliving forms of matter obey precisely the same physical laws.

Assumption B. Living states of matter are distinguishable from nonliving states of matter only by the potential for evolution, i.e. the hereditary transmission of naturally selected traits.

To make these assumptions more plausible let us consider for a moment the antithetical assumptions. Suppose, for example, that the difference between living and nonliving matter depended upon different initial conditions. From the point of view of the physicist we would have to call this a 'special creation' which may be allowable as a highly unlikely event or a miracle; but this would nevertheless be scientifically barren since it can be neither derived from any physical theory nor tested by any real experiment [6]. Furthermore, if we assumed that living and nonliving matter obey different elementary laws of motion, then, by the physicist's meaning of a law, there must be observable or derivable regularities or correlations between detailed measurements involving one type of matter but not the other. Since an enormous number of observations have been made and no such regularities have been found, this antithetical assumption seems unjustified. Notice that Assumption A does not imply that all aspects of physical theory have been formulated, but only that whatever theories we currently accept must apply equally to living as well as nonliving matter. Finally, if we reject Assumption B and assume antithetically that living and nonliving matter can both evolve in some sense, then we have only succeeded in generating a new question: Why did living matter distinguish itself by evolving so much more variety thah nonliving matter? In other words, we must have in addition to Assumption A, which states the *similarity* of nonliving and living matter, a second assumption which clearly *distinguishes* living from nonliving matter. To omit the second type of assumption is to miss the whole problem.

Accepting Assumptions A and B for our discussion, what can we conclude from them? Some physicists feel that such assumptions are contradictory. Wigner's [7] argument that self-replication is impossible, assuming only the normal laws of quantum mechanics, would fall into this category. Other physicists propose that autonomous biological laws must exist. Such arguments have been given by Bohr [8], Elsasser [9], and Burgers [10], for example.

My own point of view is that there is no scientific value whatever in attempts to dismiss such arguments because they have their basis more in the language or logic of physics rather than in the details of molecular biology. Assumptions A and B are statements of a crucial paradox which must be zealously and carefully pursued if we are to have a physical theory of general biology. Furthermore, I believe there is reason to expect that these assumptions are closely related to the central epistemological paradox of the mind-body problem itself. However, in this paper I shall emphasize this paradox only in the context of the origin of life problem. First, I shall try to clarify these assumptions so as to sharpen the paradox. Otherwise the central problem can too easily become obscured by the many details of new experimental discoveries.

▸ *What are the physical laws?* The Assumption A is relatively easy to amplify because the meaning of initial conditions and laws of physics have already been deeply analysed [11]. What we wish to emphasize, however, is that the physicists meaning of 'obeying the laws of motion' is a rigorous statement which can be quantitatively verified by measurement and calculation. An elementary law of motion is a prescription for correlating the values of certain variables which give the state of a system at any one time to the values of these variables for any other time. In this language, once the complete state of a given system has been chosen by assigning initial conditions for one time, any additional information about an earlier or later state of the system is redundant. That is, no better prediction about the future or past of the system can be made, in principle, by supplying more information. The rules for applying this descriptive language are precisely formulated and one cannot, for example, say that a molecule obeys these elementary dynamical laws of physics simply by looking at numbers representing the *average* structure of a large collection of these molecules or by moving around a desk-top classical model of one of these molecules. In other words, to say that an enzyme or nucleic acid molecule obeys the dynamical equation of motion of quantum mechanics cannot be regarded by the physicist as a justifiable conclusion without some evidence to actually support such statements.

We have therefore labelled our statement A as an assumption, because although it might be argued that quantum mechanics has in the past described correctly many diverse molecular effects, we must also consider the arguments that have been presented showing that quantum mechanics is not consistent with the basic property of self-replication.

In the clarification of Assumption B we encounter another type of difficulty. Few biologists would dispute that the living states of matter evolve by a different process than the nonliving states. In fact, the potential for hereditary evolution may be used as a definition of present life. But it might be argued that hereditary evolution is not the most elementary or fundamental condition for the origin of life. For example, simple autocatalysis, metabolism, or replicating processes may also be called primeval features of the living state. However, to be brief, I shall simply define as a necessary condition for the origin and persistence of life the property of reliable hereditary transmission of naturally selected traits. Unfortunately this phrase is not yet in the language of physics, and its meaning is often imprecise even in biology. Therefore let us try to define what hereditary transmission and natural selection can mean in the language of physics.

▸ *What is heredity?* The traditional idea of a hereditary process involves the transmission from parent to offspring of particular traits. By a trait the biologist does not mean an invariant of the equations of motion, but one property chosen from a set of possible alternatives. The trait which is actually transmitted depends upon a *description* of the trait recorded or remembered from some earlier time. Thus, the central biological aspect of hereditary evolution is that the process of natural selection operates on the actual traits or phenotypes and not on the particular description of this phenotype in the memory storage which is usually called the gene. This is essential biologically because it allows the internal description or memory to exist as a kind of virtual state which is isolated for a finite lifetime, usually at least the generation time, from the direct interaction which the phenotype must continuously face.

The crucial logical point of hereditary propagation which corresponds to the biological distinction between genotype and phenotype is that hereditary propagation involves a *description* or *code* and therefore must require a *classification* of alternatives and not simply the operation of the inexorable physical laws of motion on a set of initial conditions. As we stated in the last section, these laws of motion tell us how to transform the state of a system at a given time into the state at any other time in a unique and definite way. The equations of motion are therefore said to perform a one-to-one mapping, or more specifically, a group transformation of the states of a system. On the other hand, the hereditary process which must transmit a particular trait from a larger set of alternatives must perform a classification process, and this involves a many-to-one mapping. It is for this reason that concepts such as memory, description, and code which are fundamental in hereditary language are not directly expressible in terms of elementary physical laws. Direct copying processes, such as crystal growth or complementary base pairing in DNA, do not involve a code or classification of alternatives; and therefore, even in classical language, simple template copying processes are not a sufficient condition for evolution by natural selection. When there is no distinction between genotype and phenotype or between the description of a trait and the trait itself or, in other words, when there is no coding process which connects the description by a many-to-one mapping with what is described, then there can be no process of hereditary evolution by natural selection.

The logical aspects of this fundamental evolutionary principle were understood by von Neumann [12] in his design of a self-replicating automaton based on the Turing machine. It is significant that von Neumann's self-replicating automaton has the same basic logic that is now known to exist in cells, even though his replicating automaton was designed without any knowledge of the details of the cellular translation code and the roles of nucleic acids and enzymes. Nevertheless it was clear to von Neumann that simple template replication or copying in itself was of no interest in either the logical or the evolutionary sense, and that only a concept of heredity which includes a code could provide growth of complexity that had any real significance for learning and evolution. Thus it may be said that a threshold of logical complexity exists for the origin of evolving hereditary structures. Following von Neumann's work many papers have pursued the interesting and essential logic of this problem [13]. It is remarkable how few biologists are aware of this work and of the logical basis for a coding process in hereditary biological evolution.

▸ *The central problem.* We have now given some idea why the elementary laws of physics do not seem directly suitable for describing hereditary behaviour. At the logical level we may say that the laws of physics describe a one-to-one mapping process, whereas hereditary propagation requires a many-to-one mapping process. Or in more physical terms we may say that the elementary physical laws are symmetric with respect to time, whereas hereditary propagation requires a direction to time. Or in other words, the temporal relation between the memory of a trait and a trait itself is not symmetric.

There is of course a broad general theory of physics called thermodynamics which is capable of treating irreversible phenomena. We may therefore ask if thermodynamic or statistical mechanical theories cannot be applied to hereditary phenomena. The answer is that of course they can be applied, but they do not lead us to expect biological evolution. In fact, it is the second law of thermodynamics which at first sight appears to be the antithesis of biological evolution leading as it does to complete disorder as opposed to the increasing complexity of biological organisms. We may therefore say that the problem of describing hereditary processes in terms of the laws of physics must not only overcome the difficulty in deriving irreversible phenomena from reversible laws, but in addition it must also show how the consequences of hereditary irreversibility lead to the phenomenon of evolution in living matter rather than the complete thermodynamic equilibrium of nonliving matter.

▸ *The classical evasion of the central problem.* One popular concept of living matter which seems to evade this paradox is the so-called automata description of molecular biology. This description treats the cell as a classical machine which behaves very much like a modern large-scale computer [14]. Such classical machines clearly exhibit the property of memory storage and hereditary transmission as well as coding and classification processes. How are such classical machines described in terms of the laws of physics?

This can be done only by the introduction of a certain type of structure which controls to some extent the dynamic motion of the system, but which is not derivable directly from the basic equations of motion. In order to exhibit the fundamental hereditary property of classification, or the selection of a trait from a larger set of alternative traits, there must be available more degrees of freedom in the static description of the machine than are available for the dynamic motion of the machine. In other words, the very concept of a memory in a hereditary system implies the existence of more freedom in the static state description than in the motion of the system, since it must be dynamically constrained so as to propagate only that particular trait which is recorded in the memory storage. Such a structure which has more degrees of freedom in its state description than in its dynamic motion is called in classical physics a *non-holonomic* constraint [15]. If one accepts the classical description of non-holonomic constraints, it is possible to tailor a machine to represent almost any code or logical function which one can imagine, and this is the basis of all computer design. In fact, it is possible to program large-scale digital computers to imitate macromolecular processes in living cells, including DNA replication, transcription, and coding into protein enzymes [16]. We therefore must raise the question: Are classical descriptions or models of living cells an adequate basis for understanding the fundamental nature of living matter and its evolution?

A part of the answer to this question was already suggested by the physicist Schrödinger [17] in his book *What is Life?* which appeared in 1944. Schrödinger pointed out that the order which we associate with classical mechanisms is based on the averages of large numbers of molecules, whereas the order in the cell is based on single molecules. Schrödinger suggested that the relative stability of individual molecules can be understood in terms of the stationary states of quantum mechanical systems, but he did not discuss the transmission of this order into macroscopic systems, that is, the expression of this order as a hereditary trait. This is another statement of the central problem which still must be solved.

In order to present the problem in more detail, let us return to the classical concept of an hereditary system which must involve a non-holonomic constraint. What are some of the basic properties of non-holonomic systems? The idea of a constraint is entirely classical, arising from the treatment of some degrees of freedom as purely geometrical structures which do not depend on time and the laws of motion. However, when we look at matter in more detail, we realize that all macroscopic structures must ultimately be represented by elementary forces which hold them together. We may then distinguish permanent structures as only metastable configurations with relatively long relaxation times compared to our time of observation. For example, an ordinary clock which may, during short intervals, appear to be telling very accurate time will, over longer intervals, slowly lose this accuracy and gradually approach irreversibly the equilibrium to which all classical machines must tend. A good clock is simply a mechanical device which manages to measure the same time interval a large number of times before it reaches equilibrium. Thus at least two widely differing relaxation time scales are necessary for the description of hereditary behaviour in statistical systems, and at least one of these time scales must describe an irreversible process. Usually one of these time scales is so long that it is neglected in the treatment of the dynamical problem, and it is replaced only by geometric constraints. The more complete mathematical description of this classical hereditary behaviour in non-equilibrium, non-linear statistical mechanical systems can become very elaborate [18]. But, as Schrödinger pointed out in the case of hereditary storage, the peculiarity of biological chemistry is that all its hereditary processes are based on the dynamics of individual molecules and not on statistical averages of vast numbers of molecules. Therefore we must try to extend these classical and statistical mechanical ideas of a hereditary process to individual reactions at the quantum mechanical level.

But in view of the obvious difficulty of such a microscopic description we may again raise the question: Why is it necessary to use quantum mechanical description when it is known that in many cases, even in chemistry, a classical description is adequate for a good understanding of the processes involved? In other words, why is it not possible to admit that a quantum mechanical description would indeed be more accurate, but that for all practical purposes a classical description is close enough?

▸ *The reliability condition for evolution.* Now we have asked the crucial question: When is a theory or a description 'close enough'? We have asked this question about our own attempts at describing living matter in terms of physical laws; but certainly the same question can be applied to the hereditary process itself, and we may ask: When is the description of a hereditary trait 'close enough'? This is a very practical type of question, and its answer depends upon what purpose one has in mind for a particular theory or hereditary description. In the context of the origin of life we may restate this question as follows: When is hereditary storage and transmission reliable enough to achieve the persistent evolution of complexity in the face of thermodynamic errors, that is, in the face of the second law of thermodynamics? Even though we do not understand the mechanism, the only conclusion I have been able to justify is that living matter has distinguished itself from nonliving matter by its ability to achieve greater reliability in its molecular hereditary storage and transmission processes than is obtainable in any thermodynamic or classical system.

Now while it is reasonable to assume that the relatively high reliability of hereditary *storage* in cells is based upon the quantum mechanical stationary states of single molecules, we must still find an explanation for the relatively high reliability of the *expression* of these hereditary descriptions as classical traits which interact with the classical environment. In other words, we may say that the description of the trait is quantum mechanical, whereas the natural selection takes place on the classical level between the phenotype and the environment. But even though we do not understand the hereditary transmission process, the answer to our question whether classical laws are 'close enough' for a theory of life is now obvious; for if the cell itself cannot use a classical description for its hereditary processes, then how could we expect to describe this unique biological reliability only in terms of classical description?

We must next ask what type of physical theory can be used to describe the expression of a quantum mechanical hereditary *description* as classical interactions between the phenotype and the environment. In particular, by what physical theory do we describe the hereditary transmission process which decodes the quantum mechanical description to produce the classical phenotypic expression?

▸ *The quantum theory of measurement.* There are a few other types of phenomena in physics in which quantum and classical descriptions must be closely related – ferromagnetism, low-temperature phenomena such as superconductivity and superfluidity, and the measurement process in quantum mechanical systems. It is significant that for all these types of phenomena there exists no complete description in terms of elementary quantum mechanical equations of motion. For this reason, while it does not appear likely that an explanation of molecular hereditary transmission will be produced forthwith, at least the problem is not entirely foreign to physics. Therefore while I cannot support the optimistic belief of many molecular biologists that heredity is simple and has now been explained in terms of physics, neither can I be as pessimistic as some physicists in their assertion that living states of matter cannot be derived from physical laws.

The problem of describing a measurement process in terms of the quantum equations of motion has evaded clarification since the formulation of quantum mechanics. Since there are many papers which discuss the problem in detail [19], I shall do no more here than suggest how molecular hereditary processes are related to the quantum theory of measurement. The basic problem may be stated in the following way: The quantum equations of motion operate on unobservable wave functions which may be interpreted as probability amplitudes. Under certain conditions, these unobservable probability amplitudes can be correlated with observable variables in the normal classical world, and when this happens we can say that a quantum mechanical measurement has been executed. However, the quantum equations of motion do not appear to account for this correlation of probability amplitudes with the observable probabilities in the classical world, and a second type of transformation called 'the reduction of the wave function' must be used to produce a measurable quantity. The quantum equations of motion are reversible in time and perform a one-to-one transformation of the wave functions, whereas the reduction of the wave function or measurement is an irreversible process and involves a classification of alternatives or a many-to-one transformation. This necessity for two modes of description is at the root of the wave-particle duality, the uncertainty principle, and the idea of the necessity of complementarity in the complete description of quantum events.

However, it is also this duality which leads to the conceptual difficulties of measurement processes, since there is as yet no objective procedure for specifying where in a chain of events a measurement occurs. In other words, whether or not a measurement is said to occur depends somewhat arbitrarily on where the observer chooses to separate his quantum mechanical and classical descriptions of a given measurement situation. If he chooses to consider the entire system, including what he would normally call the measuring instrument, as only a single quantum mechanical system, then he could recognize no measurement. In the same way, if he chooses to treat a collection of molecules which includes what he normally would call a hereditary memory as only a single quantum mechanical system, then he could recognize no hereditary process [20].

▸ *Enzymes as measuring molecules*. In view of the unsatisfactory state of the theory of measurement in quantum mechanics, it is a remarkable fact that physicists continue to make accurate measurements, just as biologists continue to replicate, without, in a sense, understanding what they are doing. However, in the case of physicists this can be partially explained by the size of measuring devices, which are usually large enough to be clearly recognized and treated only as classical systems. In any case, measuring devices are designed by men and are not considered as spontaneous collections of matter. On the other hand, we cannot make this excuse for biological replication. When we speak of individual molecular hereditary transmission as similar to a measurement process, we must ask what corresponds to the measuring instrument at this microscopic level. Or in terms of the origin of life, what is the simplest molecular configuration which could express an hereditary trait and which we could have expected as a reasonable spontaneous molecular organization?

Here we must return to our fundamental definition of heredity as a classification process rather than as simple copying, or the propagation of an invariant of the motion. We have pointed out that a classical physical representation of a classification process must depend on non-holonomic constraints, that is, on structures which allow more degrees of freedom in the state description than is available for the actual dynamic motion of the system. At the molecular level this would imply that non-holonomic constraints allow a larger number of energetically possible reactions than the number of reactions which are actually available to the dynamics of the system. Now in chemical terms, reactions which are *available* as distinct from those which are energetically possible can differ only in the activation energy and entropy, so that we are led to associate the classification process or hereditary propagation with the control of rates of specific types of chemical reactions. Of course, in cells the control of rates and specificity is accomplished by the enzyme molecules. Furthermore, it is significant that classical models of enzyme mechanisms depend upon flexible structures or allosteric [21] and induced-fit [22] descriptions which are equivalent to the physicists' non-holonomic constraints. It is of course possible that other molecules such as nucleic acids also exhibit non-holonomic, catalytic properties, but this remains to be demonstrated.

As we have already noted, the physicist may design and perform experiments on quantum mechanical systems without microscopic analysis of the process of measurement, since in most cases a distinction between the quantum system being measured and the classical measuring device can be clearly specified or recognized. In other words, we accept the non-holonomic constraints of a clock, a switch, or gate mechanism because these are large classical devices with many degrees of freedom which we can statistically tailor to approximate our needs with the desired precision or reliability. But at the microscopic level it is by no means obvious that we could design a single molecule which performs with the speed and reliability observed for specific enzyme-controlled reactions. In the first place, the very idea of a non-holonomic constraint in an elementary quantum mechanical system forces on us a profound modification of the language [23]. Not only would the idea of measurement have to be extended to include non-observed quantities, but also the equations of motion are effectively modified by non-holonomic conditions, since there is no possibility in deriving such exact constraints by taking into account additional existing degrees of freedom. On the other hand, this requirement of a reliable microscopic non-holonomic constraint is consistent with the early suggestion of London [24], and more recent suggestion of Little [25], that macromolecules could conceivably possess superfluid or superconductive states which would allow change of shape or transfer of matter with no dissipation. As London pointed out, such a quantum fluid state would combine the characteristic stability of stationary states with the possibility of dynamic motion isolated from thermal agitation. This is precisely what would appear to be essential for specific catalysts which act as precise molecular measuring devices.

A direct experimental test of such a measurement theory of specific catalysis may run into a type of difficulty foreseen by Bohr, namely that external measurements of crucial life processes may be incompatible with the results of the process. If measurements by single enzyme molecules depend upon the internal correlation of their electrons, then any device which can be said to perform an external measurement on these electrons will necessarily destroy some of these correlations with the result that specificity and catalytic power of the enzyme will be correspondingly decreased. However it is not clear that other more indirect evidence may not be obtained to test such a theory [26].

It is to be expected, of course, that classical description will indeed be useful at many points, and that for many practical applications the details of the quantum mechanical description are unnecessary. However, in terms of any general theory of biological systems the *reliability* of hereditary transmission or the speed and accuracy of measurement is crucial. For example, the difference between a mutation rate of 10^{-4} and 10^{-8} per elementary hereditary transmission may easily be the difference between the immediate extinction or long evolution of a species, and no one could claim that this is a trivial difference [27]. It is this quantitative difference in the speed and reliability of hereditary transmission for which quantum mechanics can account and for which classical theory cannot.

In terms of the origin of life problem, this assumption also leads us to believe that *life began with a catalytic coding process at the individual molecular level*, since no spontaneous thermodynamic system or classical machine appears to provide the necessary speed and reliability for such a distinctive evolutionary process within the classical environment. Therefore, although with great effort we may design complicated classical hereditary machines which may adapt themselves to a classical environment for a limited time, we would not expect such complex devices to arise spontaneously on the primitive earth, nor could we expect them to achieve a statistical reliability in their hereditary processes which would allow them to distinguish themselves so successfully from the environment for five billion years.

▸ *Design of origin of life experiments*. What type of abiological experiments does this measurement theory of hereditary processes suggest? First of all we are led to believe that specific catalytic molecules are essential for the coding process in hereditary transmission. Contrary to the so-called central dogma which states that nucleic acids transmit all hereditary information and that proteins can only receive it, we would have to conclude that while template molecules or holonomic structures may be said to *store* hereditary information it is only the non-holonomic or allosteric catalysts which can *transmit* hereditary information. Moreover, it is important to realize that *a definition of stored information itself cannot usefully be made without a complete specification of the coding mechanism for transmitting it.* Without complete specification of the transmission code there is no way to determine what variables of a given physical structure consist of hereditary information which is to be transmitted, and what variables are simply to be treated as initial conditions needed to specify the storage structure at a given time. Failure to recognize that prior specification of the transmission code is necessary in order to define stored information in an objective way has led to much confusion in the use of the information concept, particularly in biological systems.

The experimental approach suggested by this theory contrasts sharply with the strategy of most so-called 'chemical evolution' or abiogenic organic synthesis experiments which emphasize the growth of non-hereditary chemical complexity as judged by the similarity of particular spontaneous species of molecule with existing biochemical species in cells [28]. While it may be relatively easy to compare the similarity of these spontaneous molecules with the evolved molecules of cells, the question of the significance of each type of molecule is left open. This has generated much discussion as to which type of synthesis is most closely related to the origin of life on earth and elsewhere. Since widely different sets of initial conditions can produce many of the same organic molecules, there have also arisen controversies over such uncertainties as the equilibrium conditions and free energy sources which actually produced the first prebiological molecules on the earth, and what extraterrestrial conditions might favour the occurrence of certain types of prebiological molecules.

I would like to point out that from the hereditary point of view it makes little difference for the general origin of life problem whether a molecule is made by heat, ultraviolet, ionizing particles, or for that matter obtained from a chemical supply house, *as long as the molecule has no memory*. Furthermore, since we can associate hereditary transmission only with rate control processes, or, in other words, since equilibrium states can have no memory, we should not expect equilibrium conditions to play a primary role in the origin of life. Of course I do not mean that organic syntheses and equilibrium considerations are not important for the origin of life problem. What I wish to emphasize is that the hereditary property itself is the only context from which these other questions can have any objective biological interpretation. Our theory therefore constrains us to look for the simplest possible hereditary chemical reaction *processes* before we can usefully compare our chemical products with living cells.

▶ *Examples of hereditary copolymer reactions*. How shall we experimentally recognize the most primitive hereditary reactions or codes in simple molecules? This is a very difficult question which I cannot fully define, but the general idea can be illustrated by a series of examples of polymer growth. Consider first a simple growing homopolymer in which there is an initial monomer addition rate constant, K_a. After the chain grows long enough, suppose that it folds into a helical conformation, say, with five monomers per turn, and that because of the folding the monomer addition rate increases to $K_a' > K_a$. The nature of the bond is not changed, only the rate has increased. One case of such conformation-dependent catalysis occurs in the N-carboxyanhydride synthesis of polypeptides [29]. The significant aspect of this simple conformation-dependent, rate-controlled reaction is that the oldest exposed monomer in a helical chain is controlling the rate of addition of the next monomer. This amounts to a delay in the control mechanism corresponding to one turn in the helix. Now this *delayed control process* may not appear to have much evolutionary potential. However, we shall show how natural modifications of such *conformation-dependent specific catalytic effects* may produce elaborate hereditary coding in simple copolymers.

Next consider a copolymer growth in which the initial comonomer addition rates are K_a and K_b. Suppose that this chain also folds into a helix with five monomers per turn and that in this configuration the proximity of the $(n-4)$th to the $(n+1)$st position catalyses the next addition step as in the previous example. However, now when we are using two types of monomer it is generally unlikely that the catalytic effect of the $(n-4)$th position is independent of the type of monomer at that position. If we now assume that there is a very strong rate-controlling effect of only the $(n-4)$th monomer on the addition of the next monomer, there will then be four possible control schemes or *codes* as shown in Table 1.

TABLE I

Code	*Monomer type in $(n-4)$th position*	*Catalysed monomer type in $(n+)$st position*
1	a	a
	b	b
2	a	b
	b	a
3	a	a
	b	a
4	a	b
	b	b

What will be the effect of these possible codes on the sequences in the copolymer chain? The last two codes will clearly degenerate into simple homopolymers no matter what the starting sequence may be. However, the first two codes will lead, respectively, to eight and four species of periodic copolymer. It is also clear that the linear sequence in each of these species is completely determined for a given code of Table 1 by any five adjacent monomers in a helical turn, and therefore each turn of the helix can be considered as a genetic sequence. For example, if an *a* or a *b* monomer at the $(n-4)^{th}$ position increases the relative rate of addition of the same type of monomer as shown in the first code of Table 1, then any of the five cyclic permutation sequences, *ababa*, *babaa*, *abaab*, *baaba*, and *aabab* are equivalent genetic sequences for one of the species. The other seven species are generated from the two homopolymers, *aaaaa* and *bbbbb*, and the sequences *babab*, *aabaa*, *bbaba*, *baaab*, and *abbba* or one of their cyclic permutations. It is important to realize that the specificity or relative catalytic power of the $(n-4)^{th}$ monomer, or in other words the *reliability* of the tactic catalyst with respect to the types of added monomer will determine the inherent rate of mutation in this type of hereditary propagation. Of course, the addition of an uncatalysed monomer, that is, the addition of a non-coded monomer, will not necessarily lead to a new species, since all cyclic permutations of the end-turn sequence are genetically redundant. This would correspond to a mutation in DNA which still codes for the same amino acid.

Suppose now that we wish to increase the reliability of such a coding process. In other words, we wish to increase the specificity and corresponding catalytic power for the addition of particular monomers. One reasonable mechanism for accomplishing this is to assume that more monomers must play a role at the active site, or in other words, that there are more interactions with the monomer which is to be added. Using the same basic model of a helical copolymer, suppose that not only the $(n-4)^{th}$ position monomer determines the type of addition but that the last monomer or n^{th} position also influences the specificity. This is sterically reasonable, since the nth and the $(n-4)^{th}$ monomer form a step dislocation in the helix at the position where the next monomer will be added. But now instead of only four possible coding schemes as shown in Table 1 there are sixteen possible codes, again assuming only absolute specificity or so-called eutactic control. If we choose the code which catalyses the addition of an *a*-type monomer when the nth and $(n-4)^{th}$ monomer are the same type and a *b*-type monomer when the nth and $(n-4)^{th}$ monomer are a different type, we will obtain four species of copolymer which may be represented by the four periodic sequences given below:

S_1: $(a)_n$
S_2: $(bba)_n$
S_3: $(bbbaaba)_n$
S_4: $(bbbbbababaabbaaabaaaa)_n$

The molecules within each species S_2, S_3, and S_4 will differ from each other only in the phase of the starting sequence. The sum of the length of all periods is $2^5 = 32$, and therefore no other eutactic species are possible for this given conformation and code. Of course, we may also specify each species by five consecutive monomers from any part of each chain. For example, S_1 : *aaaaa*, S_2 : *abbab*, S_3 : *baaba*, S_4 : *bbbbb*. It is clear that species S_2, S_3 and S_4 have two, six, and twenty other equally good starting genetic pentamer sequences respectively.

If one forms a state-transition matrix for this polymer growth process listing all thirty-two initial and final states, the hereditary property will be apparent by the reducibility of this matrix into four sub-matrices corresponding to the four species of the chain. From this state-transition matrix description it will be obvious that the growth space for a given initial five-monomer chain is less than the physically possible state space for the five-monomer chains. The mechanism for this growth process, which we have not specified here, is therefore equivalent to a non-holonomic constraint.

Of course, these simplified copolymer models are only to illustrate in the simplest way how true hereditary processes can arise at the molecular level. It is unlikely that tactic polypeptide growth would occur under so few constraints or in this particular autonomous form. The optimum conditions under which such tactic catalytic growth of polypeptides might be found on the sterile primitive earth need further discussion [30]. It is plausible from the known tactic processes in present cells, and the assumption of continuity in evolution, that the most primitive polypeptide tactic catalysis also involved polynucleotides and the constraints of particle or membrane-like surfaces. The origin of the nucleotide-amino acid code remains a deep mystery, but from what we have said, the answer should not be expected in template models or non-catalytic processes.

The reliability of copolymer catalysts. Even though we are not able to propose at present any detailed quantum mechanical mechanism for this type of conformation-dependent catalytic process, it is instructive to look for specific properties of such single copolymer hereditary catalysts which affect their reliability, since this property is essential for evolution. The significant characteristic of enzyme catalysis is that the specificity may be controlled only by weak bond interaction, whereas the catalysis or rate control operates only on the strong covalent bonds of the substrate. By contrast, classical machines, like clocks, use the strong bonded structures, such as the gears and escapements, to control the formation of weak bonds, that is, the frictional contacts between escapement pins and gear teeth. At the copolymer level a distinction between strong and weak bonds is already implicit in the concepts of monomer *sequence* and *conformation*, since neither of these terms could be usefully defined if only one type of bond strength existed between monomers. The linear sequence is in fact defined as the monomer order obtained by following the strong bonds from one end of the chain to the other, while the conformation in linear chains refers to the shapes held by the weak bonds as allowed by the rotation or flexibility of the strong bonds, but not by breaking strong bonds. Of course in enzymes there are covalent bonds cross-linking the chain, but the definition of a linear sequence is still recognized by the most stable strong bond path.

What is the effect of these different roles of strong and weak bond interactions on the reliability of hereditary propagation in classical and quantum mechanical systems? We have already pointed out, following Schrödinger, that the covalent bond in a copolymer chain provides an ideal static *storage* mechanism for hereditary information. However, it is no less important that all dynamic hereditary *transmission* processes, which include replication, transcription, and coding, operate with high reliability in the face of external and internal perturbations. In particular, it is more important that hereditary propagation cease altogether rather than propagate errors or lose the coding rules. Otherwise such uncontrolled catalytic activity only speeds up the destruction of the hereditary information. For example, in the helical copolymer model in which the helical structure is maintained only by weak bonds and the genetic memory by strong bonds we could expect some form of error prevention upon heating, since the helix will become a random coil first and thereby stop catalysing monomer addition. On the other hand, in most classical machines, such as clocks, it is more likely that upon gradual raising of the temperature the machine will begin to operate with errors before it stops altogether. In other words, unless special error-correcting devices are employed, a classical clock will tell the wrong time before it melts, whereas an enzyme will melt (denature) before it catalyses the wrong reaction. For these reasons we may expect optimum reliability and, therefore, survival value in hereditary systems in which the non-holonomic constraints representing the translation code mechanism are formed from weak-bonded structures, while the memory storage as well as the phenotypic expression of this description is preserved in strong-bonded metastable structures. Evidence of thermally inactivated specific catalysts should therefore be assigned high significance in abiogenic experiments.

However, even under optimum operating conditions there remains a certain level of random thermal disturbance which affects the speed and accuracy of any classical measuring device. Normally, when brownian motion or particle statistical fluctuations disturb the accuracy of a measurement, the only remedy is to increase the mass of the device or increase the time of observation so as to average out the fluctuations. Consequently high accuracy or precision in classical machines is incompatible with both small size and high rates of operation. We are left then with the challenging problem of interpreting the enormous speed and precision of individual enzyme molecules without being able to use the statistics of the large numbers of degrees of freedom which we associate with macroscopic objects.

At first sight such speed and accuracy in single quantum mechanical systems may appear even more difficult to explain because of the uncertainty principle. For example, we may say that if we choose to measure the energy of a system with an accuracy of ΔE, then the measurement interaction must extend over a time interval of $\Delta t \geqslant h/\Delta E$, so that speed and accuracy in this case are fundamentally incompatible. However, a more precise description of what enzymes actually accomplish does not involve such a simple relation between conjugate variables involved in the measurements. The specificity of enzymes appears to depend on the accurate fitting of a part of the substrate to a part of the enzyme. This implies that specificity depends on the measurement of relative position co-ordinates of certain regions of the substrate. But since the bond which is catalysed may be at a different location, the momentum co-ordinates conjugate to the co-ordinates determining the specificity need have no direct relation to the speed of catalysis. On the other hand, if the enzyme structure has non-holonomic properties, which we claim is necessary for hereditary transmission, this implies that dynamic correlations must exist between the measured co-ordinates determining specificity and the momentum co-ordinates involved in the catalysis. The reliability of substrate recognition and the speed of catalysis now become a problem of describing how such dynamical correlations can be maintained without invoking classical structures. As we indicated above, this is a difficult conceptual and mathematical problem.

Such reliability consideration will probably be crucially related to the size of enzymes and the structures associated with hereditary transmission, which of course includes the machinery for DNA replication and transcription as well as coding. It has been shown that the allowable accuracy of quantum mechanical measurements increases with the size of the measuring device, so that only in the classical limit can these measurements be described as exact [31]. This inaccuracy cannot be interpreted as the normal errors of measurement, or associated with the uncertainty of measuring a *pair* of non-commuting variables. Rather it is the result of the attempt to describe the measurement transformation by the quantum equations of motion. Although quantitative estimates of reliability have not been made, it is plausible that copolymers must have grown spontaneously to a certain size before they could perform tactic catalysis with sufficient reliability to assure some evolutionary success. Perhaps such reliability requires membrane- or particle-bound copolymers as found in the tactic reactions in present cells.

The main point of this discussion is to emphasize the necessity of reliable molecular coding for any persistent hereditary evolution. There are two aspects to this necessity: first, the *logical threshold* as illustrated by von Neumann (see pp. 75–7) which distinguishes the description or genotype from the construction of phenotype; and second, the *physical reliability threshold* which maintains the hereditary dynamics so that the rate of accumulation of information by natural selection can exceed the rate of error in the overall hereditary transmission process. These discussions suggest that neither template copying processes nor non-specific catalysis can account for the origin of life. Even though classical automata may be designed by man to satisfy the logical and reliability thresholds useful for a kind of hereditary evolution, we would expect that quantum mechanical description will turn out to be essential for any fundamental understanding of living matter [32]. Furthermore, the difficulties in quantum mechanical description of reliable hereditary processes do not appear to be simply a matter of complexity, but are likely to involve some of the most difficult conceptual problems which lie at the basis of physical theory. Would it be so surprising, after all, if the secret of life turned out to be based on something more than simple chemical description?

▸ *Some broader questions.* I have used the origin of life context in discussing coding and reliability because this level allows the simplest possible conception of a molecular hereditary transmission process. We have seen that even at this level the theoretical difficulties remain serious. Nevertheless I believe that the concepts of coding and reliability will not only be useful, but also crucial at all levels of biological organization – cellular, developmental, evolutionary, and certainly in the higher nervous activity associated with the brain. We have used code to mean the relation between an elementary genotype and a phenotype, that is, a relation between a physical symbolic description and the physical object which is actually contructed from this symbolic description.

The process of cellular replication and in particular the development of the organism may be interpreted as an entire system construction process which requires a coding mechanism which interprets as well as replicates a description. Largely from studying the logic of abstract automata we may begin to appreciate how, through the discovery of simple codes, it is possible to generate elaborate ordered structure from relatively concise descriptions. Such a description-code-construction process cannot be adequately characterized as either preformation or epigenesis, since on the one hand the construction may be totally unlike its description, whereas on the other hand the description and code structure together provide a complete, autonomous generation of the phenotypic construction within the crucial limits of reliability.

At the evolutionary level this concept of a symbolic genetic description and its code structures must be broadened to a larger system which includes not only the description of the system itself but also a description or a 'theory' of the environment. In the evolutionary context the phenotype itself now plays the role of a composite measuring device which tests the descriptive theory through its interactions with the real environment. In this language we must also expand the concept of reliability to include the overall predictive value of this description-code or theory-measurement system. I believe it is then reasonable to associate this overall predictive value with what is called the 'measure of fitness' in evolutionary theory.

Finally, at the level of nervous activity in the processes of memory and intellectual theory making, we are again searching for more elegant code structures which allow the maximum predictive reliability over the widest domain, but which can be generated from relatively short symbolic descriptions. Perhaps we could even say that the characteristic sign of biological activity at all levels is the existence of efficient and reliable codes. However, at none of these levels can we evade the basic question of how biological systems achieve the unique reliability of their codes through which they have so clearly distinguished themselves from nonliving matter. Even at the level of memory and consciousness it is possible that single enzymes may provide the crucial transmission links or codes from the senses to the internal descriptions in the brain.

SUMMARY

We have asked once again the historical question: Are the characteristic processes of biological organisms understandable in terms of the basic laws of physics?
I have tried to show that in spite of the many classical models of cellular structures and functions there are severe difficulties in accounting for the reliability of hereditary transmission in terms of the elementary laws of physics. I have proposed that the ultimate source of the unique distinction between living and nonliving matter does not rest on idealized classical models of macromolecules, template

replication, or metabolic control, but on the *quantitative reliability of molecular codes which can correlate the contents of a quantum mechanical description with its classical phenotypic expression*. To understand such a correlation between quantum descriptions and the corresponding observable classical event requires a quantum theory of measurement applied to elementary molecular hereditary processes. Such a theory presents serious, though I hope not insurmountable, conceptual and formal difficulties for the physicist. However, in spite of the unsolved theoretical questions we can specify certain necessary conditions for individual molecular coding structures. These conditions suggest that the seat of coding or measurement processes in living matter is the individual non-holonomic enzyme catalyst, although it is likely that other structures in the cell serve to increase the reliability of these codes.

Broadly interpreted, the existence of a molecular code of exceptional reliability is essential not only for the origin of life, but also for the development of the individual, the evolutionary process of natural selection and survival of hereditary traits, and even the symbolic coded descriptions which we call intellectual theories. But whatever level of complexity we study, we may expect to find the conformation-dependent, tactic catalyst serving as the most elementary hereditary transmission device. For these reasons I believe that describing such reliable hereditary molecular events in terms of quantum mechanics remains the fundamental problem which we must study, not only for theoretical biology, but perhaps also for a firmer epistemological basis for physical theory itself.

ACKNOWLEDGMENTS

This work is supported by the Office of Naval Research, Contract Nonr 225 (90), and the National Science Foundation, Grant GB 4121. The paper was prepared while the author was a member of the Center for Theoretical Studies, University of Miami, Coral Gables, Florida.

Notes and references

1. See, e.g. E. P. Wigner *Proc. Natl. Acad. Sci. 51* (1964) 956.
2. An example of such a statement is found in J. D. Watson *The Molecular Biology of the Gene* (W. A. Benjamin: New York 1965) 67: 'Until recently, heredity has always seemed the most mysterious of life's characteristics. The current realization that the structure of DNA already allows us to understand practically all its fundamental features at the molecular level is thus most significant. We see not only that the laws of chemistry are

sufficient for understanding protein structure, but also that they are consistent with all known hereditary phenomena.'

A much earlier, pre-DNA optimism toward heredity was expressed by Thomas Hunt Morgan in 1919: 'That the fundamental aspects of heredity should have turned out to be so extraordinarily simple supports us in the hope that nature may, after all, be entirely approachable.' (Quoted from F. H. C. Crick *On Molecules and Men* (University of Washington Press: Seattle 1966.)

3. E.g. from D. C. Phillips *Scientific American 215* (1966) 90: '. . . as a result of all the work now in progress we can be sure that the activity of Fleming's lysozyme will soon be fully understood. Best of all, it is clear that methods now exist for uncovering the secrets of enzyme action.'

4. E.g. from J. C. Kendrew *Scientific American 216*, no. 3 (1967) 142 (reviewing *Phage and the Origins of Molecular Biology*, J. Cairns, G. Stent, and J. Watson, eds.): '. . . up to the present time conventional, normal laws of physics and chemistry have been sufficient, and at least in the opinion of this reviewer the forward horizon is clear of awkward facts that will require new or paranormal laws for their explanation.'

5. See, e.g. V. M. Ingram *The Biosynthesis of Macromolecules* (W. A. Benjamin: New York 1965): M. F. Perutz *Proteins and Nucleic Acids* (Elsevier Pub. Co.: Amsterdam 1962); J. D. Watson *The Molecular Biology of the Gene* (W. A. Benjamin, Inc.: New York 1965).

6. Attempts to justify chance as an explanation of the origin of life are often made, e.g. G. Wald *Scientific American* (August 1954) 3. For a physicist's attitude toward chance as an explanation see P. W. Bridgman *Science 123* (1954) 16.

7. E. P. Wigner in *The Logic of Personal Knowledge* (Routledge and Kegan Paul: London 1961) p. 231. Also see P. T. Landsberg *Nature 203* (1964) 928.

8. N. Bohr *Atomic Physics and Human Knowledge* (John Wiley and Sons: New York 1958). See esp. pp. 21 and 101.

9. W. M. Elsasser *Atom and Organism* (Princeton Univ. Press: 1966); also *The Physical Foundations of Biology* (Pergamon Press: New York 1958).

10. J. M. Burgers *Experience and Conceptual Activity* (MIT Press: Mass. 1965).

11. See, e.g. R. M. F. Houtappel, H. Van Dam, and E. P. Wigner *Rev. Mod. Phys. 37* (1965) 595. For a shorter discussion see E. P. Wigner *Science 145* (1964) 995.

12. J. von Neumann in (L. E. Jeffress, ed.) *Cerebral Mechanisms in Behavior* (John Wiley and Sons: New York 1951) p. 1. A more technical treatment is J. von Neumann in (A. W. Burks, ed.) *The Theory of Self-replicating Systems* (Univ. of Illinois Press: Urbana 1966).

13. See, e.g. M. Arbib *Information and Control 9* (1966) 177; C. V. Lee in *Mathematical Theory of Automata* (Polytechnic Press: Brooklyn) p. 155; J. W. Thatcher in ibid, p. 165. J. Myhill in (M. D. Mesarovic, ed.) *Views on General Systems Theory* (John Wiley: New York 1964) p. 106.

14. See, e.g. J.-P. Changeaux *Scientific American 212* (1965) 36.

15. A. Sommerfeld *Mechanics* (Academic Press: New York 1952) p. 80; E. T. Whittaker *A Treatise on the Analytical Dynamics of Particles and Rigid Bodies* 4th ed. (Dover Publ.: New York 1944). ch. VIII.

16. W. R. Stahl and H. E. Goheen *J. Theoret. Biol. 5* (1963) 266; W. R. Stahl ibid. *14* (1967) 187.

17. E. Schrödinger *What is Life?* (Cambridge Univ. Press 1944)

18. See, e.g. E. Frieman and R. Goldman *J. Math. Phys.* 7 (1966) 2153 and references therein.

19. Two classical references on the subject are: W. Heisenberg *The Physical Principles of*

Quantum Theory (Dover Pub.: New York 1930) esp. ch. IV; J. von Neumann *Mathematical Foundations of Quantum Mechanics* (Princeton Univ. Press, 1955) Two recent discussions are: E. P. Wigner *Am. J. Phys. 31* (1963) 6; A. Daneri, A. Loinger, and G. M. Prosperi *Nuc. Phys. 33* (1962) 297. Broader more conceptual discussions are: N. Bohr in *Atomic Physics and Human Knowledge* (John Wiley: New York 1958) p. 32; P. K. Feyerabend and G. Süssman, each in S. Korner (ed.) *Observation and Interpretation in the Philosophy of Physics* (Dover Publ.: New York 1957).

20. This follows the most widely held interpretation as found, e.g., in von Neumann, loc. cit.

21. J. Monod, J.-P. Changeaux, and F. Jacob *J. Molec. Biol. 6* (1963) 306.

22. D. E. Koshland, Jr. *Fed. Am. Soc. Exp. Biol. Proc. 23* (1964) 719.

23. R. J. Eden, *Proc. Roy. Soc. 205A* (1951) 583.

24. F. London, *Superfluids*, 2nd ed. vol. I (Dover Publ.: New York 1961) p. 8.

25. W. A. Little *Phys. Rev. 134* (1964) A 1416.

26. For example, enzyme catalysis has been studied at very high magnetic fields ($\sim$220,000 gauss) by B. Rabinovitch, J. E. Maling and M. Weissbluth, *Biophys. J.* (in press), ibid. *7* (1967) 187. No effects were observed; however, owing to the uncertainties in the theory and the fact that critical fields are higher in small superconductors, these results by no means exclude the possibility of superconductive or superfluid properties in enzymes.

27. Although classical approximation may be useful for many types of biological description, we also expect that the problem of the speed and reliability of codes at quantum mechanical dimensions will not be limited to the evolutionary context. In particular, memory and thought in the brain appear to encounter the same type of difficulties with small size, high capacity, and reliability. But in the case of consciousness there is in addition the more obscure problem of the physical basis of self-reference.

28. A list of abiogenic synthesis experiments to 1964 can be found in H. Pattee in *Advances in Enzymology* (F. Nord, ed.) vol. 27 (John Wiley and Sons: New York 1965) p. 381.

29. M. Idelson and E. R. Blout, Polypeptides XVIII, *J. Am. Chem. Soc. 80* (1958) 2387.

30. H. H. Pattee in (A. D. Ketley, ed.) *The Sterochemistry of Macromolecules* (Marcel Dekker: New York, in press).

31. H. Araki and M. M. Yanase *Phys. Rev. 120* (1960) 622.

32. J. D. Cowan in (N. Wiener and J. P. Schade, eds.) *Prog. in Brain Research* vol. 17 (Elsevier Pub. Co.: New York 1965) p. 9.

Towards a Physical Theory of Self-organization

Karl Kornacker
Massachusetts Institute of Technology

Organization, which for the moment might be thought of as some set of structure-function relations, is a dominant feature of all biological systems and appears at all levels from the molecular to the social. The basic questions one asks about organization – what it is and how it arises – are as difficult as the same questions asked about life.

My approach to these questions has involved the search for a physical language, as distinct from a purely mathematical or purely metaphorical language, which can describe and explain the appearance of organization *at any level*. Perhaps the best way to summarize the view I have developed is to consider the relations between Maxwell's demon, thermodynamic coupling, the 'utilization of information', and biological organization.

Maxwell's demon is an imaginary device which was thought to show that the second law of thermodynamics fails when the microscopic state of a system is known. The demon, which can 'see' molecules, is able to generate concentration gradients across a boundary by operating a 'trapdoor' that allows molecules to cross in one direction only. If one ignores the entropy increase in the demon, then the entropy decrease produced in the demon's environment appears to violate the second law. When the entropy increase of the demon is taken into account, however, the demon simply becomes a model for one possible mechanism of thermodynamic coupling (compensating entropy changes in different components of an isolated system).

A Maxwell's demon 'utilizes information' because he performs a task which would be impossible without 'knowledge' of local molecular movements. It therefore seems reasonable to say in this case that *the strength of thermodynamic coupling is a measure of the utilization of information*. Since we want to use a similar definition to describe all levels of organization, however, it is important to show that these considerations are not restricted to molecular phenomena.

The general phenomena we will be interested in are regularities which 'emerge' in complex systems. For example, the selective sensitivity of a single optic nerve cell to a special visual pattern emerges from the interaction of thousands of synaptic currents generated in the dendritic tree of the cell. In turn the visual response of the organism emerges from the interaction of millions of optic nerve fibers.

Application of the concept of thermodynamic coupling to emergent regularities is straightforward once we are able to apply statistical thermodynamics to systems whose microscopic state can in fact be specified. The above discussion of Maxwell's demon bears on this point; the following brief discussion of heat suggests how the programme is carried out.

For a particle subject to an external force, F, and moving with velocity, v, the energy, E, changes according to

$$dE/dt = F \cdot v$$

If now we average this expression (over time, or over different particles, or over repeated experiments with different initial conditions) we have, using $\langle\ \rangle$ to denote the averaging operation,

$$\langle dE/dt \rangle = \langle F \cdot v \rangle$$
$$= \langle F \rangle \cdot \langle v \rangle + \langle (F - \langle F \rangle) \cdot (v - \langle v \rangle) \rangle$$

Granting that some such averaging operation is required to reduce the noise level in a meaningful macroscopic measurement, we can say that, relative to such measurements, the rate at which macroscopic work is done is $\langle F \rangle \cdot \langle v \rangle$. By the first law of thermodynamics, then, the other term, $\langle (F - \langle F \rangle) \cdot (v - \langle v \rangle) \rangle$, which measures the strength of correlation between the fluctuations of F and v, must give the rate at which energy is transferred as heat.

Returning now to the nerve cell example, the firing pattern of the cell is determined by a (non-linear) spatial and temporal average of the dendritic synaptic currents, so that spatial and temporal correlations in the pattern of dendritic activity can have a direct expression in the resulting cell response. The language of statistical thermodynamics seems appropriate for the study of these correlation phenomena, and might even prove effective for the analysis of pattern recognition.

Coming now to the definitions of organization, we will say that *functional organization at some given level is equivalent to thermodynamic coupling* (*utilization of information*) *at the same level*. It would seem also that *a structure could be called organized if its existence were either necessary for the maintenance of some functional organization, or dependent on the operation of some functional organization.* Without reference to functional organization it seems to me impossible to define structural organization in a useful way.

In conclusion we might briefly consider the origin of organization. The approach outlined here suggests that the answer may emerge from an analysis of the kinetics of thermodynamic coupling during non-stationary irreversible processes.

A Party Game Model of Biological Replication

reprinted from Nature, *Vol. 212, No. 5057, pp. 10–12, 1st October 1966*

D. Michie and C. Longuet-Higgins
University of Edinburgh

Comparisons are so commonly made between the information processing systems of the living cell and the operations of digital computers that it seems worth trying to work out at least one comparison in detail. We have taken for the simple exercise reported here the vegetative propagation of a single-celled organism – for definiteness let us say the *E. coli* bacterium. Our task, then, is to specify the minimal informational transactions required, in the manner in which a computer systems analyst might approach it. By this we do not mean something elaborate: on the contrary, simplicity and clarity are all. The following party game provides a sufficient, and certainly an entertaining, vehicle for a minimal specification.

> *Phase 1*. Prepare a form set out as in figure 1. Hand it to one of your guests. It may be wise to 'plate out' one or two further duplicates of the form, in case the first falls on barren ground. Having ensured that supplies of paper and ink are available, allow a suitable interval to elapse. Then search your guests and confiscate for analysis all documents in their possession. What will you find?

For the present we will leave this for the reader to work out for himself, since experience shows that the discipline of solving this problem is essential to prepare the mind for the next stage of the demonstration.

> *Phase 2*. Now restore to your guests all confiscated documents, having made suitable inventories of your findings. After an interval, release among them one or more copies of a smaller form, set out as in figure 2. Again allow time to elapse, and make a second harvest. What does the second harvest bring?

During phase 1 it is clear that the large forms (*E. coli*) will multiply logarithmically among the guests (nutrient niches), assuming on their part literacy and basic willingness to co-operate (adequate physical and chemical constituents). The limit to multiplication is set when either the guests are saturated with forms or the supplies of paper or ink (specific nutrients) are exhausted. Let us assume that before this stationary phase is reached we initiate phase 2, which represents infection of the *E. coli* colony with one or more bacteriophage virus particles. What happens next?

It can easily be verified that the small forms will begin to multiply within the bodies of the large forms, finally bursting the bounds of the latter and causing their destruction. This completes the 'paper model' of the multiplication of *E. coli* bacteria and their parasitization by bacteriophage particles which are replicated by courtesy

I. TRANSLATE INTO ENGLISH AND OBEY THE DIRECTIONS IN THE BOX BELOW.

(1) Remplir la plume, s'il faut.

(2) S'il y en a, prendre un grand morceau de papier, dessiner auprès de la tête une boîte à vingt-trois lignes vides, et au-dessus de cette boîte imprimer comme suit:

'I. TRANSLATE INTO ENGLISH AND OBEY THE DIRECTIONS IN THE BOX BELOW.'

Au-dessous de la boîte imprimer comme suit:

'II. IF YOU NOW POSSESS A NEW PRINTED FORM, COPY THE CONTENTS OF THE ABOVE BOX INTO THE EMPTY BOX ON THE NEW FORM. OTHERWISE, IF THERE IS AN EMPTY BOX AT THE FOOT OF THIS FORM COPY INTO IT AS MANY LINES OF THE MESSAGE AS THERE IS ROOM FOR, AND THEN RETURN TO INSTRUCTION I; OTHERWISE STOP.

'III. HAND BOTH OLD AND NEW FORMS TO OTHER PEOPLE.'

II. IF YOU NOW POSSESS A NEW PRINTED FORM, COPY THE CONTENTS OF THE ABOVE BOX INTO THE EMPTY BOX ON THE NEW FORM. OTHERWISE, IF THERE IS AN EMPTY BOX AT THE FOOT OF THIS FORM COPY INTO IT AS MANY LINES OF THE MESSAGE AS THERE IS ROOM FOR, AND THEN RETURN TO INSTRUCTION I; OTHERWISE STOP.

III. HAND BOTH OLD AND NEW FORMS TO OTHER PEOPLE.

FIGURE 1

Form to be 'plated out' among the party guests, thus initiating a logarithmically growing colony of paper organisms. The box contains the genetic program for constructing an entire new organism. This program must be translated and executed by the somatic apparatus of the cell, represented here by the English-language text outside the box. Note that faithful replication of this apparatus is achieved by specification, not by a process of copying. But the genetic program (DNA) is copied. In order to allow parasitization by virus (figure 2) we must imagine enough blank space at the foot of the above form to accommodate a hundred or more versions of form 2 of the host's own metabolic machinery. Why does the model work, and what are the precise equivalents between macromolecular structures and the elementary components of the party game?

First, we will give the equivalences not so far stated. The box in the large form represents the genetic apparatus of the *E. coli* cell. The French text contained in it corresponds to the genetic message inscribed in the DNA of the bacterial chromosome. This text has to be translated, if only transiently, into an English message before it can be executed: the equivalent process is the transcription of the DNA code-script into the corresponding representation in messenger RNA. The printed English text of the large form represents the somatic apparatus of the cell, concerned with implementing the genetic specifications of the cell (instruction I), replicating the chromosomal DNA (instruction II) and supervising the mechanics of cell division (instruction III). For these operations actually to be carried out, active co-operation from the party guests is of course required, and this may be felt to be a serious flaw in an analogy which casts the human participants in the passive role of nutrient niches for our paper organisms. It is certainly asking too much of our model to expect a close fit in every detail. Formal correspondence can, however, be preserved if we imagine our guests as mindless beings with no volition beyond a compulsion to execute any instruction which they

IF A LARGE PRINTED FORM WITH A BOX ON IT COMES INTO YOUR POSSESSION, CUT OUT THE CONTENTS OF THE BOX BELOW AND PASTE INTO THE TOP OF THE BOX ON THE LARGE FORM; THEN READ THE LARGE FORM ITSELF.

(*a*) S'il y a assez d'espace au pied de cette page, imprimer là un petit avis, qui inclut une boîte a dix lignes vides; au-dessus de cette boîte imprimer comme suit:

'IF A LARGE PRINTED FORM WITH A BOX ON IT COMES INTO YOUR POSSESSION, CUT OUT THE CONTENTS OF THE BOX BELOW AND PASTE INTO THE TOP OF THE BOX ON THE LARGE FORM; THEN READ THE LARGE FORM ITSELF.';

s'il n'y a pas assez d'espace, séparer tous les petits avis au pied de la page, distribuer les à vos amis, et détruire le résidu mutilé.

(*b*) Ne pas lire le texte qui suit dans cette boîte.

FIGURE 2

Second form, to be released when the colony initiated by the first form has become well established (see text) chance to read. Under these conditions the guests act merely as a substrate, and the cell's 'genetic box' contains complete and effective specifications, given the right somatic environment, for manufacturing an entire new cell.

Phase 2, as previously remarked, is intended to model the infection and metabolic subversion of the host bacterium by particles of virulent phage. To understand the biological interpretation of our formal scheme it is sufficient to note that, once again, genetic information is expressed in French; the printed English text which surrounds the 'genetic box' is again somatic: it corresponds to the protein shell discarded by the virus on penetration of its bacterial host and re-synthesized at a later stage by bacterial ribosomes under orders from viral DNA.

Now note one or two consequences of our logical model, which present some rather instructive features. First, suppose that the sequence of viral DNA which says '(*b*) Ne pas lire la texte qui suit dans cette boîte' were by accident damaged or corrupted. The consequence would be that the infecting phage particle, after integration into the 'genetic box' of the host, would not enjoy unrestrained replication, but would instead be replicated in an orderly way at each cell division as part of the bacterial chromosome. An attractive feature of this deduction is that it corresponds exactly with the behaviour of temperate phage. Although this detailed correspondence may be fortuitous and hence misleading, it seems justifiable to point it out. In the same spirit, we cannot refrain from imagining the consequences of a failure to supply enough ink 'pour remplir la plume'. This could correspond to inhibition of protein synthetic activity, identifying 'ink' with nitrogenous substrate: the consequence will be arrest of DNA replication.

In sum, we can say that the logical relations of the biological system concerned have been modelled to a fair approximation. If there were no more to it than that, then we would merely have learnt a new educational game. We believe, however, that our game may actually possess explanatory power. To say why we think so demands that we set up another set of equivalences, this time with computer systems rather than with living cells.

The French text now corresponds to a user's computer program written in some programming language such as FORTRAN. The transient English language translation represents the machine language version which has to be produced by the computer before the program's specifications can be carried out. The printed English text which surrounds the 'program space' in the computer store corresponds to the supervisor ('monitor', 'operating system'). The supervisor in a large computer, such as *Atlas*, is an elaborate master program which controls and allocates input and output devices, accepts, stacks, and assigns priorities to the incoming stream of user's programs, and sees to the execution of these programs, first translating each with the aid of the appropriate compiler. Thus, in our example, the supervisor arranges for the translation of the user's program from FORTRAN to machine language and initiates its execution (instruction I), makes a copy of the original program (instruction II – as when the IBM 7090 FORTRAN system gives the user a 'listing' of his program), and (in a hypothetical world in which computer software is free to spread vegetatively through a receptive network of inter-compatible computers) dumps the entire computational output into a virgin computer of the same make (instruction III).

We shall not spend much thought on the computational analogy of infection in the world of computers. This kind of trick can be played by any computer programmer who possesses sufficient knowledge of the supervisor and sufficient disregard of the sanitary rules which discourage users from deliberately 'overwriting the system'. To the extent that these rules are incorporated into the system software itself, the rogue programmer's intent to subvert must be backed by additional knowledge and ingenuity if he is to succeed. The point we wish to emphasize here – indeed, the point towards which our entire argument is oriented – is that this *segregation of user's software and system's software* has evolved among computer users for sanitary and economic reasons, and that, as Weismann first noted, a similar segregation of germ plasm (DNA specifications) and soma (cellular machinery of implementation) has evolved naturally. If we merely want a program capable of causing its own replication, in addition to its other action, why do we not use simple self-reference, as follows:

1. Do this and that.
2. Copy the whole of this message on to a fresh sheet of paper and pass both copies on to other people.

We propose that the operational objections to this are similar in the biological and the computational cases. The reason why computer scientists do not in general encourage the user to write a program which will (*a*) act on itself or (*b*) modify the supervisor during execution, arises from considerations of public order: if such transgression of boundaries were allowed, difficulties would be created for the same or other users of the same program, or of other programs using the same supervisor. The penalty of having a relatively invariant somatic apparatus is the hazard of subversion by agencies able to exploit this invariant structure. That the penalty has been found worth paying in both contexts is possibly not coincidental. The analogy seems sufficiently striking to bear further investigation.

Comments by H. H. Pattee

The authors' main point is the practical necessity of the segregation of the program and operating machinery of the computer or, in biological terms, the necessity of the separation of genome and phenome. It should be mentioned that the logic of this necessity was discussed by von Neumann in his theory of self-reproducing automata (J. von Neumann, *Theory of Self-reproducing Automata*, edited by A.W. Burks, University of Illinois Press, Urbana, 1966, pp. 121–6). Von Neumann asks essentially the same question as the authors, why bother to copy the *description* of a machine, why not just copy the machine itself? The reason he gives is similar to the authors', that a description is essentially a passive structure which may be 'explored' or measured without affecting the operation of the machine it describes, whereas measurement of the states of the active machine directly would be expected to cause unwanted changes in the machine behaviour. Von Neumann then goes on to point out that the construction of a description from the machine itself runs into logical antinomies of the Richard type. This suggestion is based on one of Gödel's theorems that the complete epistemological description of a sufficiently rich language cannot be given in the same language alone. This is the basis of theorems on undecidability (*ibid.*, pp. 47–56). Hence it is difficult to imagine the occurrence of genome and phenome function through a process of evolution which itself requires such a distinction in the first place. Thus the spontaneous occurrence of a physical description-translation process appears to be the primeval example of the chicken-egg paradox.

I would like to add that in all these discussions the actual representation of the description is assumed to be of the same physical nature as the machine itself, that is, both description and construction are classical or thermodynamic aggregations. But as I pointed out in my discussion (pp. 69–73), living matter, or what von Neumann calls 'natural automata', differs crucially from artificial automata in just this respect, namely, living systems appear to use quantum mechanical descriptions which are somehow translated into classical or thermodynamic machines or phenotypes. I believe that here any model of life based only on macroscopic artificial automata will break down, not only because the physical situation is different, but also because the *logic* of translation is different. In the case of artificial macroscopic automata the translation process may be described statistically by non-equilibrium non-linear thermodynamic correlations. The reliability of the translation will consequently increase with the size and decrease with the speed of the logical elements. The errors in such systems may be adequately treated by the normal statistical or thermodynamic theories of noise.

On the other hand, if the description of the automaton is entirely quantum mechanical while the machine is itself a classical device, then the translation process must amount to a quantum mechanical measurement. The description of noise or reliability in this situation remains a very difficult conceptual and formal problem, but one which any general theory of biological behaviour cannot evade.

Theoretical Biology and Molecular Biology

C. H. Waddington
University of Edinburgh

It is recognized by all far-sighted biologists that we are in the middle of a period of exceptionally rapid advance in our understanding of the most fundamental of biological processes, those concerned with heredity and development. It might be asked, do we need any more of 'Theoretical Biology' than is contained, or which we may reasonably hope will soon be contained, within the growing body of Molecular Biology?* If we reach an understanding of the elementary processes, is that not sufficient to allow us to feel that we understand in principle the whole gamut of biological phenomena? The argument might be reinforced by a reference to physics. It seems at first sight that theoretical physics is concerned only with whatever is most fundamental or elementary at a particular period – the Bohr-Rutherford atom at one time, quantum physics a little later, intra-nuclear particles at present. The main endeavour seems always to be to deepen the comprehension of ultimate elements, the relation between these elements and complex physical systems such as macromolecular chemistry, solid state physics, etc., being left to specialists who can almost be dismissed as 'mere engineers'.

Can theoretical biology similarly concentrate its attention on the elementary processes studied by molecular biology, and leave all the rest to the lesser breeds of natural historians, ecologists, physiologists, metabolic biochemists, and the like? There are several molecular biologists who seem tempted to answer Yes, and there are many more who in actual discussion behave as though no statement can be a satisfactory answer unless it is phrased in chemical terms – which implies that no question is a real question unless it also is posed chemically. This point of view is, unfortunately, not represented in the essays collected in this book. But, even in the absence of an explicit defence of it, it seems worth mentioning a few of the arguments against it.

* To avoid misunderstanding, let me emphasize that the point under discussion here is not the old 'vitalist-mechanist' catch-as-catch-can wrestling match – nowadays usually referred to in America as the conflict of reductionists versus anti-reductionists. I am taking it that all externally observable biological phenomena – a category which *excludes* conscious awareness – are ultimately explicable in terms of concepts confluent with those used in the physical sciences, remembering that the physical sciences are themselves open-ended, so that even the most doctrinaire reductionist cannot tell biologists just what they have to reduce their systems to. The question here might be paraphrased as: Do you have to wait till you can reduce to Molecular Biology of the Dogma in a single leap, or is there anything useful to do meantime?

In the first place, it should be pointed out that the analogy between theoretical biology and theoretical physics involves a comparison of two bodies of theory which are at very different stages of historical development. The theory of the physical world already contains very well worked out accounts of the connections between highly complex systems – whether artefacts such as jet engines or transistor radios, or natural phenomena such as thermal elasticity, reactions of chemical radicals, etc. – and underlying units which are tolerably elementary. Thus any refinement in our understanding of these units can, in most cases, easily be incorporated into a well-established structure of theory which deals with the more complex end of the spectrum of physical reality. In contrast, general theories relating the more complex biological phenomena to elementary units are not very satisfactory. The most important of these theories for biology are those of evolution and of epigenesis. During the last few decades, when the most elementary units known to biologists were genes, these theories did not become as thoroughly worked and solidly established as did, for instance, the theories of chemistry during the period in which its fundamental units were atoms with valencies. Thus when the chemists' atom turned into a quantized wave function, the theoretical physicists could feel confident that the chemists had a sound theory into which to incorporate the new ideas. But when the 'old-fashioned gene' turns into a replicating sequence of bases in DNA which is active in controlling a DNA-RNA-protein sequence, theoretical biologists have little reason for any confidence that there are in existence sound theories of evolution and development waiting to be enriched, rather than thrown into chaos, by the new insight.

Biology is in fact still in the process of creating its theories 'from the elementary units to the complex', at the same time as it is making such rapid advances in the analysis of the units. But the analysis by itself is not sufficient until the intellectual apparatus for building upwards towards the complex is better developed.

Molecular biologists, like most people, feel little surprise at hearing criticism of others. They tend to react to the points made above by saying: 'Of course we know that general biological theory is in a poor way; it will just have to wait until we can provide it with the right answers – which will be molecular answers.'

But this final point is not really such a bold claim as it sounds, and as it is often intended to be. Of course the final answers to biological problems must, ultimately, be in molecular terms. What, other than molecules, is there for biological systems to be constructed out of? The important point about the answers is not whether they are in molecular terms or not, but whether they are answers to the important questions. It can only be in relation to general theoretical biology that a molecular biologist can hope to guess which of the many problems he can envisage are likely to be of strategic rather than merely tactical importance.

Consider some examples. It is obviously a strategic point – one which began to be envisaged some thirty years ago – that the genetic material is organized in a linear fashion down to the level of small-molecular units; what is the importance, for a fundamental theoretical biology which might be applicable on Mars, of the fact that these units are nucleotides rather than amino-acid residues, which was the first guess of most biologists? Again, theory would suggest that any system which can evolve by natural selection would require both a rather stable, and therefore unreactive, memory store, and a phenotype more ready to react with the environment in ways giving rise to natural selection; is it a mere tactical 'happen-so' that life on earth has settled on DNA for the former role and proteins for the latter, with peculiar intermediate links of RNA in between – just what is the significance of altering this sugar?

The adoption of a system in which the memory-store is chemically different to the reactors involves a coding relation between them. Again one can ask if the actual code is merely happen-so, or whether it is fully or partially determined by an internal logic. An answer to this, which may not be long in coming, will certainly be a major contribution to the basic theory of biology. So was the suggestion of the idea, which was soon shown to be true, that the hereditary material takes the form of a duplex of complementary linear aggregates. Even if this had remained for some time a purely abstract notion – that is, if biologists had not been able to identify the chemical nature of the complementary strands – it would have greatly clarified our thinking about the process of genetic replication. Of course its importance was still further increased by the fact-finding exercise which showed that the substance involved is a double helix of DNA. But without prior recognition and formulation of the logical structure of the processes of gene replication and gene action, no one would have known in which direction to search for relevant facts, nor appreciated their significance when found.

It is rather generally felt that one of the next major biological problems which molecular biology will, or should, clear up is that of cellular differentiation in higher organisms. There is, apparently, a promising lead provided by Jacob and Monod's clarification of the control of enzyme synthesis in bacteria. But it requires careful consideration – and this would be a task for theoretical biology – how far the analogy between induced enzyme synthesis in bacteria and higher-organism differentiation is adequate. On the one hand, the empirical facts to be explained look as though they may be irreducibly more complex than the simply on-off switching (even with quantitative control) which occurs in bacteria. In the best analysed instances of cell differentiation there seem to be three phases, which may well all be necessary steps; firstly, the appearance of 'competence', or readiness to enter the next step – a transient condition, not definitely known to be cell-heritable; secondly, 'determination', or being switched into a definite path of differentiation – a condition known to be inheritable through many cell generations, with only occasional 'transdeterminations' into some other path, but not *known* to be dependent on a chromosomal mechanism; and thirdly and finally, activation, which leads to the actual synthesis of the cell-specific proteins [1].

Looking at the problem from the other side, there are many more possible types of control process than merely one which decides when a given stretch of DNA would produce mRNA. The possibility of translational, as well as transcriptional, control is of course well recognized – indeed, it was suggested well before the DNA-RNA-protein scheme was generally accepted [2]. But this does not anything like exhaust the list of possibilities. As an example, consider a preliminary and, at first sight, innocent-looking step, the replication of the DNA. The general rate of replication of the whole genome may be controlled, for instance in Gurdon's transplantations of brain nuclei into newly fertilized amphibian eggs [3] or in Dipteran salivary gland cells infected with a microsporidian parasite [4] – and it would be a bold man who asserted that these two controls operate through the same mechanism. Again there is accumulating evidence that different DNA-RNA polymerases act at different rates, and possibly differentially on different stretches of DNA; and the DNA-DNA replicases might also vary in a similar fashion. Methylation of the DNA may alter its affinity for a given polymerase. Again, there may be control of quite a different kind, involving the amplification of particular stretches of DNA. We know that the nucleolar organizer region of the chromosome in Xenopus and Drosophila contains many apparently identical repetitions of cistrons coding for ribosomal RNA [5]. This amplification is 'genetic' in the sense that it is transmitted to all the cells of the organism, and genomes of related species have different degrees of amplification. There can also be 'epigenetic' amplification in which more copies of particular DNA stretches are produced in certain cells. This certainly affects the nucleolar DNA in amphibian oocytes, in which very numerous nucleoli, each containing a 'core' of DNA, are produced and thrown off from the chromosome [6], a process which has been shown to be under feedback control from other parts of the cell [7]. Such amplification control has so far only been demonstrated for the nucleolar cistrons, but there is no obvious reason why it should not occur much more widely during differentiation. Determination might, for instance, involve the amplification of certain stretches of DNA, which did not start producing mRNA until 'activated' perhaps many cell generations later.

These are just a few examples to illustrate the wide range of control systems which may act on the DNA; and there are probably others. In the whole cellular mechanism of differentiation DNA is only one part, and perhaps a rather inert one at that. We have to consider, at a minimum, the DNA, the chromosomal proteins, the mRNA, the ribosomes, the activating enzymes, the pools of amino-acids, the transfer-RNAs, and the various replicases, polymerases, degrading enzymes, enzymes attaching or detaching mRNA and DNA, or mRNA and ribosomes, and so on. It is the task of experimental molecular biology to discover examples of differentiation controls operating by effects on one or other of these constituents. But even when one or two good examples are described – for instance, if we discover just what is happening when chromosome puffing is 'activated' by ecdysone [8] or ion-balance [9], we shall not be very much wiser until the facts can be fitted into a more fully worked out theoretical framework. It would be more illuminating, because more unexpected, if it could be shown that some conceivable type of control actually does not occur in fact – as, one dares to hope, it may be demonstrated that no changes in the DNA-protein code are found throughout the whole realm of nature.

It seems to me that we need, in the first place, a logically exhaustive classification of all the conceivable types of control. This is certainly not available at present, and may be difficult to provide. Possibly one might start from the consideration that each component (the DNA, mRNA, etc.) must (i) be made out of pool of specific sub-units, (ii) by a specific production agent (enzyme), (iii) in a particular quantity, (iv) with a certain half-life; to produce an effect it must (v) interact with a component earlier in the sequence (e.g. mRNA with ribosomes), and (vi) give rise to a product which acts 'later' in the sequence (e.g. mRNA with ribosomes gives proteins). We have therefore at least six controllable factors in relation to each of the classes of components of which some eight or nine were mentioned in the last paragraph. It is probable that not all the modes of control could even in theory operate on all the constituents, but even so the conceivable control mechanisms are at least some fifty to sixty in number; and if we considered all possible modalities of specificity there might well turn out to be many more.

The next step in a theoretical approach would be to try to deduce which of this plethora of possibilities would seem most likely to be of general importance. Useful control mechanisms must be reliable, repeatable, often heritable, must usually control batteries of genes rather than single ones, must usually give rise to sharply distinct cell states, and one could probably think of other necessary properties. From such considerations it should be possible to get some useful hints of what it would be most profitable to look for. We do not yet have such a worked-out theory. When I first started thinking about differentiation in terms of molecular control mechanisms I plumped for a mechanism which was, perhaps, the most obvious one in the pre-Crick era, namely control through alterations in the pools of amino-acids or nucleotide precursors [10], and I wasted a good deal of time playing with unnatural amino-acids and purines or pyrimidines. I think we already know enough to realize that this type of control, though it may well operate in certain cases, is unlikely to be one of the major systems on which cellular differentiation depends. It is one of the tasks of theoretical biology in this field to clarify and codify considerations of this kind. There will probably turn out to be two or three really basic and fundamental control mechanisms, probably of more sophisticated kinds than those envisaged in orthodox 'advanced' molecular biological thought of today. After all, evolution has had a long time to cook up some really clever tricks! Just to go blinding ahead on the basis that it's all a question of gene activation, and histones are trumps, may probably lead to some intriguing but equivocal results, of good temporary sales value, but is unlikely, in my opinion, to get to the heart, or should one say hearts, of the matter.

Notes and References

1. For a slightly fuller account of these phases see C. H. Waddington *Principles of Development and Differentiation* (Macmillan: New York 1966) and *Proc. Roy. Soc., B, 164* (1966), 219.

2. C. H. Waddington in Kitching (ed.), *Recent Development in Cell Physiology* (Butterworth: 1954), p. 105.

3. J. B. Gurdon *Endeavour 25* (1966), 95.

4. C. Pavan and R. Basile *Science, 134* (1966), 1956.

5. F. M. Ritossa and S. Spiegelman *Proc. Nat. Acad. Sci., 53* (1963). 737; M. L. Birnstiel *et al., Nat. Cancer. Inst. Monog., 23* (1966), 431.

6. H. G. Callan *J. Cell Sci., 1* (1966), 85.

7. M. L. Birnstiel and E. Perkowska. Personal communication, 1967.

8. U. Clever *Chromosoma, 17* (1965), 309.

9. H. Kroeger and M. Lezzi *Ann. Rev. Entomol., 11* (1966), 1.

10. C. H. Waddington *Symp. Soc. Exp. Biol., 2* (1948), 143.

A Note on Evolution and Changes in the Quantity of Genetic Information

C. H. Waddington University of Edinburgh
R. C. Lewontin University of Chicago

Evolutionary progress has often been discussed in terms of the acquisition of new genetic information. Now the quantity of genetic information contained in a genome can be assessed in the number of nucleotide pairs, i.e. total mass of D N A. There is, in fact, no clear-cut relation between this quantity and evolutionary advance. It is true that metazoa in general contain a considerably larger quantity of D N A than do bacteria. Within the metazoa, however, D N A per cell varies from species to species in a manner which shows little relation to the phylo-genetic position of these species. The largest quantities of D N A per genome are found not in the most highly advanced organisms, but, for instance, among vertebrates, in some of the amphibia. There are grounds for thinking that much of this variation may be connected with 'amplification', i.e. the incorporation into the genome of many copies on a single locus, a topic as yet very little understood.

During informal discussions over cocktails at the Villa Serbelloni an argument was advanced by Waddington, and given a more rigorous statement by Lewontin, which seems worth recording (it might be given the name 'The Serbelloni Theorem').

It states that any tendency to increase the quantity of information in the genome during evolution will be held in check because the rate of advance under natural selection will be inversely proportional to the number of informational units.

This was initially advanced on the intuitional grounds that it is always easier to achieve a definite goal when handling a relatively small number of items than when handling a crowd. A more formal argument is as follows:

Suppose the total range possible for a character is from 0 to 2A and further suppose that there are *n* identical gene loci contributing to the character. For simplicity of demonstration let us also assume the genes act additively both between loci and between alleles at each locus. Then each locus has effects 0, *a*, and *2a* for the homozygote *bb*, heterozygote *Bb*, and homozygote *BB* respectively. With this notation the additive genetic variance at any locus is $\sigma^2_{Gi} = a^2 p_i q_i$

But by hypothesis $a = \frac{A}{n}$ so that

$\sigma^2_{Gi} = \frac{A^2}{n^2} p_i q_i$ and the total additive genetic variance over all loci is

$$\sigma_G = \sum_i \frac{A^2}{n^2} p_i q_i = \frac{1}{n} A^2 (\bar{p}\bar{q} - \sigma^2_p)$$

We see then that for a given total phenotypic range A the additive genetic variance is inversely proportional to the number of loci determining the character. Also, it decreases for a fixed average gene frequency when there is a large variance in gene frequency from locus to locus. Since, by Fisher's Fundamental Theorem of Natural Selection, the rate of advance under selection is equal to the additive variance, our intuitive argument is correct.

Another way to look at it is that any intermediate value of a character can be achieved without additive variance by accumulation of some loci homozygous *B* and some homozygous *b* in just the right proportion. If *p* loci are fixed at *b* and *q* loci at *B*, then $\sigma^2_{\bar{p}} = pq = \bar{p}\bar{q}$ and there is obviously no genetic variance for selection to operate on.

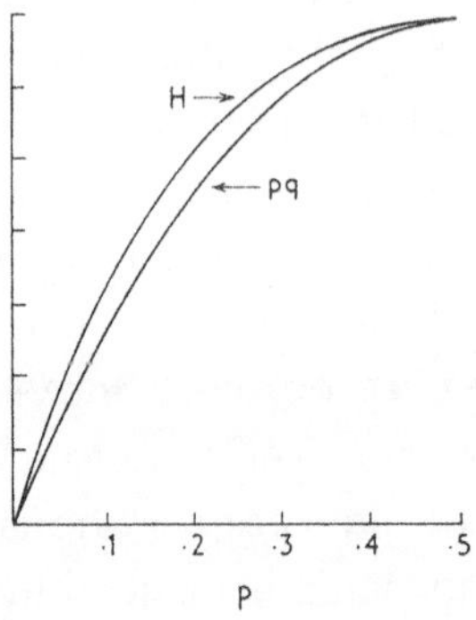

FIGURE 1

When we turn to the question of the amount of information contained in the genome there is some difficulty of definition. If we regard the genome as a sentence in which the letters are allelic states of each gene, we can use the Shannon definition of information $H_i = -[p_i \ln p_i + q_i \ln q_i]$ where p_i and q_i are the frequencies of the alternate alleles at the *i*th locus as before. But the product $p_i q_i$ used in the previous formulation of variance is very close to H in form. Figure 1 shows the product p_i and q_i and H plotted together and rescaled so that they have the same value at $p = 0{\cdot}5$. Thus we can rewrite the formula for genetic variance as

$$\sigma^2_{Gi} = \frac{kA^2}{n^2} Hi$$

where k is an arbitrary constant. As a result total genetic variance can be written as

$$\sigma^2_G = \frac{kA^2}{n} \bar{H}$$

so that if the average information per gene is fixed the rate of change of the population under natural selection decreases as the information is spread out over more and more genes.

Does Evolution Depend on Random Search?

C. H. Waddington
University of Edinburgh

Most people who consider the theory of biology at the present time seem firmly wedded to the notion that the essential process of evolution is dependent on random search. To give one example: Pattee in a paper 'Physical Theories, Automata and the Origin of Life' (in *Natural Automata and Useful Simulations*, Spartan Books, Washington; Macmillan, London; p. 76) gives one section the heading 'Is evolutionary theory a satisfactory model?' and writes: 'Evolutionary theory is based on the panacean phrase "random search and natural selection in mutable, self-replicating populations" which is used to account, in principle, for the development of almost any degree of molecular intricacy without direct regard for the physical laws which govern the elementary events. It should be borne in mind that the only direct experimental evidence for this evolutionary process is derived entirely from highly evolved organisms of relatively recent and local ancestry.'

As Pattee's sub-heading suggests, many physical scientists find difficulty in accepting random search as an adequate mechanism. The difficulties were strongly voiced at a symposium held at the Wistar Institute in April 1966 (*Mathematical challenges to the neo-Darwinian interpretation of evolution*, Wistar Institute Symposium monograph No. 5, 1967). For instance, Murray Eden referred to the negligible chance that 'a child arranging at random a printer's supply of letters would compose the first 20 lines of Virgil's Aeneid'. He goes on to offer some numerical estimates of numbers and probabilities considered by him to be relevant to evolution. 'Let us consider first the space of polypeptide chains of length 250 or less. We may think of words which are 250 letters long, constructed from an alphabet of 20 different letters. There are about 20^{250} such words, or about 10^{325}.' He then computes that the total number of protein molecules that has ever existed is of the order of 10^{52}. This is sufficient to indicate the general trend of his argument. Again Schützenberger writes (p. 74): 'According to molecular biology, we have a space of objects (genotypes) endowed with nothing more than typographic topology. These objects correspond (by individual development) with the members of a second space having another topology (that of concrete physico-chemical systems in the real world). Neo-Darwinism asserts that it is conceivable without anything further, that selection based upon the structure of the second space brings a statistically adapted drift, when random changes are performed in the first space in accordance with its own structure. We believe that this is not conceivable.'

Before discussing the arguments for and against the adequacy of random search as a mechanism of evolution it is as well to enquire whether in fact modern biology does inevitably come to the conclusion that the directions of evolutionary change are brought about by random search controlled by natural selection.

One argument tending in this direction is a theoretical one. The variations on which natural selections act have their ultimate origin in alterations of the genetic material. These alterations may take the form of changes in the sequence of nucleotides in the DNA, additions or removals of nucleotides, etc. For present purposes we do not need to look any further into the generally accepted principle that these primary genetic changes can be regarded as random mutations. Does it follow that the variations on which natural selection acts are also random? The pebbles forming the gravel on a river bed have their form determined by random processes, i.e. are the results of random search; it does *not* follow that random search plays any important part in the erection of a bridge built of concrete made out of this gravel. The factors that have to be considered in the engineering of the bridge belong, we may say, to a different order of complexity to those involved in the formation of the aggregate of which the concrete is constituted. We need to ask whether the attribution of evolution to natural search does not depend on a similar confusion of different orders of complexity – an error very similar in logical type to Whitehead's well-known 'Fallacy of Misplaced Concreteness'. The question can best be considered in two stages: first, the evolution of higher organisms, as exemplified for instance in the lineages of horses and their relatives, which have been so well studied throughout the Tertiary period; second, evolution at the molecular level, as seen in haemoglobin, insulin, etc., and as, we might argue, it must have proceeded during the origin of life.

In higher organisms it seems fairly clear that the changes which are evolutionary successful are not in general dependent on single-gene mutations resulting from random mutation. The great majority of random gene mutations which produce effects marked enough to be individually identifiable turn out to be harmful, and to be eliminated by natural selection. There are, of course, some exceptions, such as melanism in Lepidoptera in industrial areas; but even there the incorporation of a major gene mutation into the evolutionary sequence involves a simultaneous selection of a large associated group of 'modifying genes'. In general, however, the evolution of higher organisms depends on the selection of characteristics which are more or less equally influenced by large numbers of genes, i.e. they are comparable to blocks of concrete rather than to individual pebbles. What an evolving population has to do, in fact, is to produce, in some way or other, a phenotype which can find succcess, in the heterogeneous circumstances of the world around it, in leaving more offspring than other competing phenotypes. As far as its success in natural selection is concerned, it makes no difference whether or not environmental factors operating during early stages of its life history have made a large contribution to the phenotype. In practice, higher organisms have nearly all evolved rather efficient mechanisms for adjusting their phenotypes in a useful manner to surrounding circumstances ('learning' in a very broad sense). Thus physiological and developmental adaptation does usually play a considerable part in determining the phenotype.

In the production of a phenotype efficient at some task important to natural selection, such as, in horses, running fast to escape from enemies, or being able to feed on tough or low-growing herbage, very many genes will be concerned. Going back to the analogy with concrete, one might say that the role of random processes is to produce a gravel which can form a concrete capable of setting into a decent cast of an introduced object, which corresponds to the environment. The cast will not reproduce the details of the object if the pebbles in the concrete are too large, but it will not have adequate strength if they are too small. What is required is some optimum mix. There is nothing but the random process of mutation to produce the pebbles; but this is very far indeed from the conclusion that the casts formed around different objects – i.e. the phenotypes in different environments – are also produced by random search.

It is perhaps worth while to make this point more concretely in terms of a definite example. I will quote one in which I myself have worked with at least some of the genes that might be involved in an evolutionary change. Let us suppose that some environmental change puts a natural selective pressure on a population of Drosophila to increase the speed of its flight. Now, simplifying rather drastically, the flight of a fly is brought about by rapid periodic contractions of a series of muscles running between the upper and lower sides of its thorax (Fig. 1). These change the curvature of the flexible body wall, to which the wing is attached by a complex three-dimensional link which is designed in such a way that, as the body wall changes shape, the wing moves up and down. The forward motion of the insect through the air depends both on the shape of the wing-stroke, which is in some more or less figure-of-8-shaped path, and on the exact aerodynamic properties of the wing blade. A speeding up of flight could be achieved by modifying various parts of this mechanism. We know mutant genes, with developmental effects strong enough to make them useful tools in a laboratory, which affect several of the components. Our present knowledge of population genetics would suggest that other weaker-acting alleles of these genes are almost certain to be present in any wild population, and therefore available for natural selection to utilize.

For instance, alteration in the amplitude of the wing-beat or in the exact path followed by the wing-tip might be produced by altering the mechanical or geometrical properties of the surface of the thorax. This can be done by genes

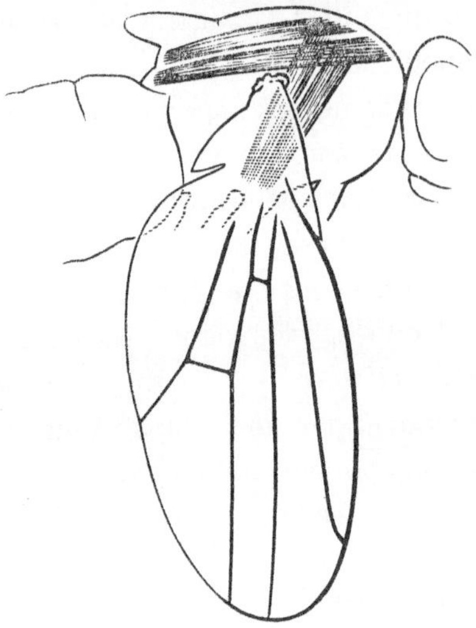

FIGURE 1
Simplified diagram of the flight apparatus of Drosophila

such as *dumpy, humpy, thick, thickoid, dachs, dachsous,* etc. Alternatively, the properties of the muscles might be altered, or the supplies of enzymes or substrates which provide the energy resources. Not much is known about genes affecting those parts of the system, but they certainly must exist. Again we know little about genes effecting the wing attachment structure. On the other hand, we know a great many genes affecting wing size, shape and aerodynamic properties. The overall size can, for instance, be altered by genetic systems affecting the number of cells composing the wing, or by different systems affecting the size those cells attain. We know both single genes with strong effects (such as *miniature, dusky,* which reduce the size of the wing cells), and also complex polygenic systems affecting cell size and cell numbers, such as those studied by Forbes Robertson. Again, there are genes affecting the shape of the wing outline, and they can do this in at least two different ways : some, such as *broad* and *narrow*, affect the rates of cell divisions orientated along the length or across the width of the wing. Others, such as *dumpy* and *lanceolate*, affect the manner in which the wing contracts from a fat inflated condition which it has at one period in its development. Such changes in wing shape certainly affect the aerodynamic properties of the wing. These could also be altered by changing the flexibility of its different regions. The leading edge of the wing is stiffened with a thick vein, and there are four other veins running from the base to the tip of the wing, with two cross-veins between them, but the posterior margin of the wing remains very flexible. Now we know a whole battery of genes, operating in several different ways, which can affect the pattern of this venation and thus the general flexibility of the wing-blade. For instance, *veinlet* removes all the tips of the longitudinal veins. *Radius incompletus* removes only the second longitudinal vein ; *cupitus interruptis* produces breaks in the fourth and fifth veins ; *cross-veinless* removes the posterior cross-vein, and so on. Finally there is another group of genes such as *Curly, curved,* and many others which cause the whole wing-blade to be more saucer-shaped, with a concavity either upwards or downwards. All these could affect the aerodynamic properties of the wing and thus the efficiency of the flight mechanism.

It is from this large array of possibilities that natural selection has to find some solution or other that brings about an increase in the speed of the insect's flight. This is a very different task from that envisaged in the phrase 'random search', which is usually taken to imply that the population has to wait for a new gene to turn up which produces by itself a more or less specific effect which natural selection is demanding. This is perhaps the natural conclusion to draw if one takes the mathematical treatments of Fisher and Haldane as profound contributions to our conceptual understanding of evolution, rather than as drastically simplified first steps towards expressing our theories – which are actually far more complex and subtle – in mathematical terms. A reading of the other great pioneer in the mathematical expression of evolutionary theory, Sewell Wright, would have warned the unsuspecting physicist to tread a little more warily.

If one attempts to diagram the situation in two dimensions, the random search model discussed by Eden might be represented as in Figure 2 (the space in which the random walk takes place is of course really multidimensional). The system that really operates is much more like that represented in Figure 3. In this we have

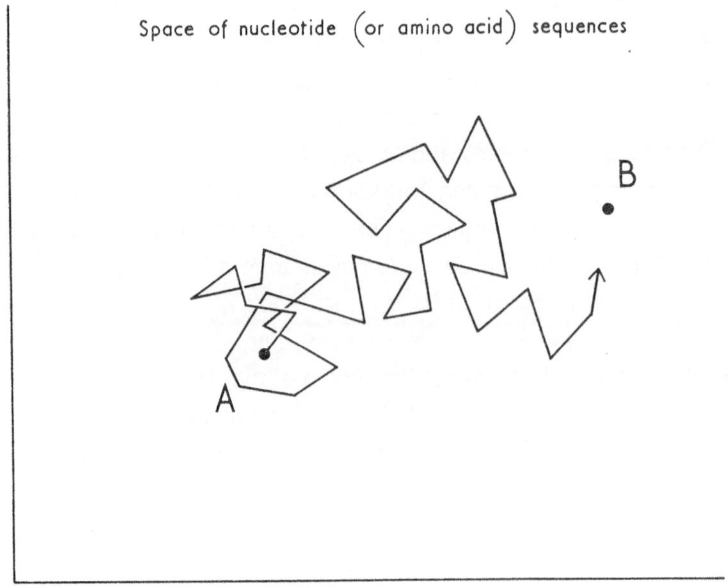

FIGURE 2

Illustration of a random search process, in which a system is required to move from point A to point B by random changes in single sub-units

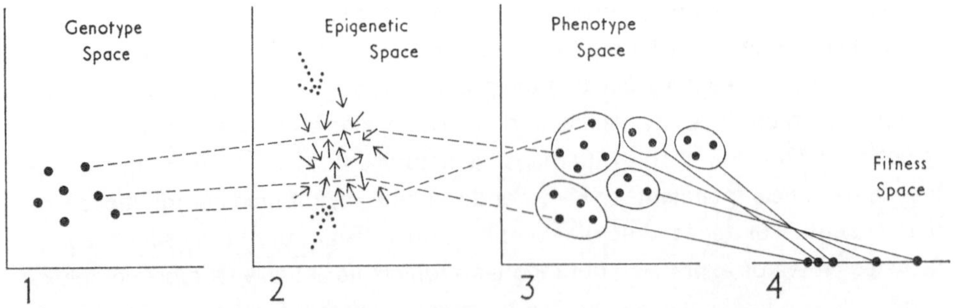

FIGURE 3

Illustration of the process of natural selection in higher organisms. We start with a population of genotypes in a multidimensional 'genotype space' (1). These are mapped, through a multidimensional space of epigenetic operators (2) (operators arising from the environment are suggested by dotted arrows), into an also multidimensional space of phenotypes (3). This is then mapped, by some complex function, into an essentially one-dimensional 'fitness space' (4) in which the only variable is the coefficient of fitness, or Malthusian parameter, i.e. the number of offspring produced. It is in the fitness space that natural selection acts; but fitnesses which are within a tolerance limit of one another are in practice indistinguishable a multidimensional space of genotypes which is mapped through a multi-dimensional space of epigenetic operators into another multidimensional space of phenotypes; this in turn is mapped down into a 'fitness space'. This last is essentially one-dimensional, being constituted only by quantitative variations in a single fitness coefficient or Malthusian Parameter. However, the mapping from the phenotype space into the fitness space is a 'tolerance mapping' in the sense of Zeeman (see for instance 'Tolerance Spaces in the Brain' in this volume). That is to say, natural selection cannot effectively distinguish between phenotypes whose fitnesses are too close together (for a more sophisticated discussion of this point see William Bossert 'Mathematical optimisation: Are there abstract limits on natural selection', in the Wistar Institute Symposium on *Mathematical Challenges*? etc., cited above).

When we turn to consider the evolution of molecules, the case against random search is weaker, but by no means negligible. It has, of course, to be quite a different case. We are dealing here with the situation of one gene, one protein, and there is no possibility of arguing that random mutations do no more than produce an assemblage of genes related to the evolutionary product as the pebbles in concrete are related to a bridge. The problem we have to consider is: What are the processes by which evolution produces an enzymatically effective protein? Now the biochemical effectiveness of a protein does not, in general, depend directly on its primary structure, i.e. the sequence of amino acid residues. It depends on the shape and properties of the tertiary, or higher order, configuration into which the amino acid chain becomes coiled. This configuration is determined in large measure by the primary structure, but the general recognition, which is becoming more and more widely accepted, that proteins may exhibit allosteric properties in relation to other molecules in their environment, shows that the primary structure does not entirely determine biochemical properties of the protein, without reference to anything else. Thus even a single protein molecule has a 'phenotype', in the characterization of which the environment plays a part.

The effectiveness (= natural selective value) of a protein depends mainly on one or a few reactive sites on the surface of the tertiary configuration into which the protein becomes folded. These sites involve only a small fraction of the whole amino acid chain. The rest can be regarded as packing or scaffolding, and have not much more to do with the activity of the protein than had the scaffolding which allowed Leonardo da Vinci to lie on his back in a convenient position to paint the ceiling of the Sistine Chapel to do with the resulting art-work.

The sequence of amino acids in a protein can be altered not only by changing one nucleotide pair in the DNA – which is presumably a completely random process. There are also possibilities of rearranging, duplicating, or inverting whole sequences of amino acids, comprising considerable numbers of units, by processes such as intra-cistronic recombination (cf. the ideas of Smithies on the determination of antibodies). These processes can also be considered 'random', but in a sense quite different from that which would be implied if one was thinking only of a 'random walk' through the space of all possible sequences of an amino acid chain of a given length. Moreover, as Fraser pointed out at the same Wistar Symposium (p. 90), as soon as organisms reach an evolutionary level at which they have diploid genomes, the possibilities of recombination enormously increase the efficiency with which random processes can explore a space of possible variations. Utilizing this, he has written computer programmes which lead him to assert that 'the genetic mechanism is a superb learning device', in contrast to the biologically less sophisticated computerers, such as Murray Eden, who conclude that 'every attempt to provide for "computer" learning by random variation in some aspect of the program and by selection has been spectacularly unsuccessful' (p. 11).

The fundamental problems of molecular evolution seem to me to be these:

1. We have somehow to get a protein which folds up into a tertiary structure which possesses a site which to some extent 'fits', and therefore has enzymatic activity on, a certain substrate. The properties of this site may be influenced not only by the primary amino acid sequence, but also by other molecules in the environment, among which the substrate molecules themselves may be important.
2. The next steps in evolution involve improving and stabilizing the effective characteristics of these sites.

For process 2 evolution could pursue either of the two courses which are open to it in the evolution of higher and more complex entities; it could either improve the adaptability of the molecule, i.e. the ease with which the properties of the reactive site could be altered by the substrate itself, so as to render the site more effective; or, secondly, it could pursue the strategy of canalization, i.e. elaborate the rest of the amino acid chain in such a way that it stabilizes an effective reactive site, independent of the presence of the substrate molecule. To perform these tasks there are available processes which are random at two levels; firstly, random changes of single nucleotides, and secondly, rearrangements of sequences of nucleotides.

For the first of the evolutionary steps – getting a protein molecule which has *some* biochemical activity of a useful kind – the same two types of mutational process are available, but I have the feeling that the former must be more important in this connection.

Until we know more about the nature of active sites on enzymes, how detailed their specification has to be, and how much they can exhibit adaptation to the substrate molecules, it seems impossible to estimate whether these random processes would have a reasonable chance, firstly, to bring effective protein molecules into being, and secondly to improve and stabilize them into the types of proteins which we meet at the present day. It is, however, only in this area that the adequacy of random search as a basic evolutionary mechanism presents a problem. As argued in the first section of these notes, we certainly do not have to suppose that a vertebrate eye, the leg of a horse, or the neck of a giraffe is in any important sense the result of random search.

The Counting Problem

J. Maynard Smith
University of Sussex

Development seems to occur in a stepwise fashion, the completion of each step setting the stage for the next. Each step may be comparatively simple, even if the end result of a long sequence of such steps is exceedingly complicated. One way in which a theoretical approach may be helpful is in identifying characteristic steps (of which there may be comparatively few kinds), and in suggesting possible mechanisms for them. This is the approach adopted in this paper; it was also adopted by Wolpert in his contribution to the conference.

The particular developmental step considered is the generation of a constant number of similar parts. For example, how is it that most men develop five fingers on each hand and 29 precaudal vertebrae? This problem, which I shall call the counting problem, resembles Wolpert's problem of the French Flag in that it illustrates in a simple form many of the characteristics of morphogenesis. In discussing it I shall first discuss what kinds of 'counting machine' could exist, and then ask to which kind embryos belong.

By a 'counting machine' I mean a machine which can do one of two things. It can either do something a fixed number N times when it is stimulated, or do something when it has been stimulated N times. A reversible machine which could do one of these things could do both, and even for irreversible machines there is likely to be a structural similarity between the two types. Embryos are of the former type; strictly, a single egg only develops five fingers on the right hand once, but a set of similar eggs each develops five fingers, which is equivalent to a single egg developing five fingers every time it is fertilized. Animals with nervous systems may be of either type, i.e. they may perform an action when they have been stimulated a fixed number of times, or they may respond a fixed number of times to a single stimulus.

Strictly, such machines ought to be called 'number generating machines', and the phrase 'counting machines' should be reserved for those which can do something a variable number of times (the number varying predictably according to the nature of the stimulus received) or which respond in qualitatively different ways according to the number of times they have been stimulated.

There would seem to be two main kinds of counting machine, which I will call ratio counters and digital counters. A simple ratio counter is shown in the figure.

Suppose that the volume of water required to fill the upper vessel is n times that required to fill the lower one from A to B, and that when the lower vessel is almost full it tips the seesaw. Then if n lies between 5 and 6, the seesaw will rock five times if the upper vessel is filled. There are two points to note about this machine, which appear to be general properties of ratio counters. First, there has to be some 'quantizing mechanism' – in this case supplied by a siphon with a flow rate which is rapid compared to the rate of flow from the upper vessel – which will break up a continuous process into a number of discrete events. Second, the number n

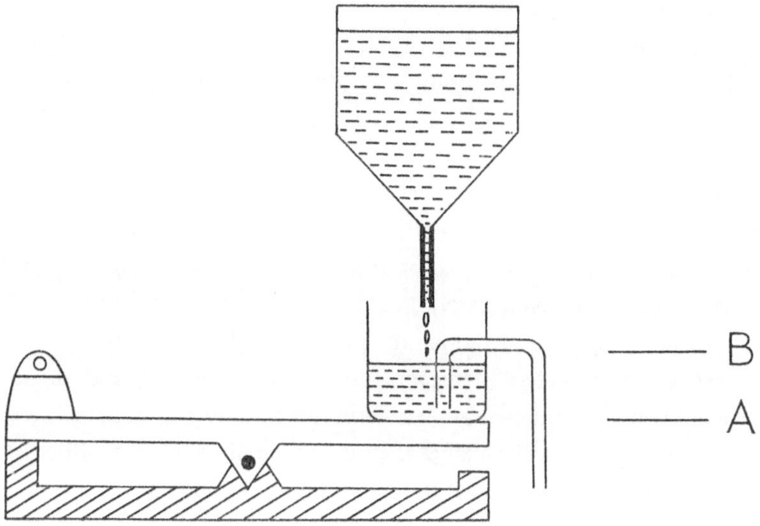

generated depends on the ratio between two continuous variables; if n is to be constant, then the range of variation of these variables must be restricted.

In digital counting, the 'machine' must again be able to generate discrete events, but the number of events is regulated by pairing them off one by one with a pre-existing set of entities contained in the machine. These entities may be all alike. For example, in cricket a bowler delivers six successive balls and is then replaced by a second bowler. The umpire often counts these deliveries by starting with six pebbles in his left-hand pocket, and after each delivery passing one pebble to his right-hand pocket. When he has moved the last pebble, he calls 'over', and the second bowler takes over.

In other cases the pre-existing entities may be unalike, and connected in sequence by rules. Thus, an umpire who wished to place six pebbles in his pocket might take pebbles one at a time from a pile and place them in his pocket, saying as he did so, 'one – two – three – four – five – six'. To do this he must have in his brain before he starts 'representations' of six qualitatively different numbers, together with a set of counting rules – e.g. after saying four, say five – and a rule saying, 'when you have said six, stop'. This is a mechanism which can easily be modified (by varying the last rule) to generate different numbers in response to different inputs, i.e. it is a counting machine proper rather than merely a number-generator.

Notice that in digital counting the counting machinery is separate from the quantizing mechanism; thus in the first example the umpire with his pebbles is the counting machine, but he is not responsible for the fact that bowlers deliver balls one by one and not continuously.

At first sight it might seem that there is no difficulty in explaining how embryos count. For example, an organism which required to produce a polypeptide consisting of 163 phenylalanine residues could do so by having a length of DNA of which the relevant strand consisted of 489 adenine bases. But this is not where the difficulty lies. Suppose for example an animal had a constant number of 163 segments (to the best of my knowledge, no animal achieves such a feat), the difficulty lies in explaining how the ability to produce a polypeptide 163 residues long could be translated during development into 163 segments. In this respect the counting problem is typical of morphogenetic problems; thus molecular biology is able to explain the shape of macromolecules – or it will be able to do so when the way in which amino acid sequence determines tertiary structure is understood. But it does not seem very likely that the *shapes* of cells or of higher organisms are determined by the shapes of their constituent molecules, as the shape of a jigsaw is determined by the shape of its pieces, or the shape of a virus may be determined by the shape of its constituent molecules.

Mechanisms of ratio counting in embryos are not difficult to think of. One possible mechanism was suggested by Turing [1]. Normally, when a number of reacting chemical substances are free to diffuse throughout some region, or 'morphogenetic field', they will reach a stable equilibrium distribution such that their concentrations are uniform over the field. Turing showed that for certain values of the reaction and diffusion rates this uniform equilibrium is unstable, and any small disturbance will lead to a standing wave of concentrations, with a 'chemical wavelength' λ separating the peaks of concentration. If one of these substances induced cells exposed to it to differentiate in a particular way – e.g. to form a bristle – such a process could account for the regular spacing of structures on a surface. This mechanism can be made somewhat more plausible by allowing for recent knowledge about the control of protein synthesis [2].

The development of a chemical wave provides a quantizing mechanism. If a wave develops within a field of given size, a counting mechanism is also provided. Thus in the one-dimensional case, if waves of chemical wavelength λ develop in a field of length s, then, since an integral number of waves must be formed, the number actually formed is typically the nearest integer to s/λ.

If for some reason the size of the morphogenetic field is changed, then we should expect a change in the number of structures formed. This is often the case -- for example, Kroeger has described how the number of wing veins in Ephestia can be altered by altering the size of the field. Nevertheless, difficulties do arise in trying to explain the constancy of numbers of repeated parts by ratio counting. For example, if the number 30 is to be generated, with an error rate of 5% or less (and vertebrates, annelids and arthropods all do as well as this in determining segment number), then the ratio s/λ would have to have a coefficient of variation of less than 1%. A review of the coefficient of variation of some macroscopic linear dimensions in animals [3] suggests that 5% is a low value. Thus the accuracy with which segment number is determined is something of a paradox. It does not really help to say that dimensions on a molecular scale can be determined very accurately, since this leaves a problem of translation.

One possible solution is a process of multiplication. Suppose that instead of generating the number 30 in one step, the body were first divided into five parts, and each of these were then divided into six parts; the total number 30 could be accurately generated without excessive accuracy in the individual steps. There is a simple way of spotting this if it happens. Variants, if they do occur, will differ from the norm not by one but by some larger number. It is interesting therefore that the segment number in centipedes varies in twos. But I know of little other evidence for processes of multiplication. Perhaps the most illuminating feature of the 'multiplication' model is that it provides what I believe is the most important reason why development is stepwise and hierarchical: there is a limit to the complexity of the patterns which can reliably be generated in one step.

Digital counting may be involved when a number shows great constancy both in evolution and development. Some examples of digital counting are trivial. For example, the number of spinal nerves is determined by the number of somites by digital counting. But this is uninteresting, mainly because it does not explain how microscopic information is translated into macroscopic, i.e. how the specification of macromolecules specifies the large-scale structure. A more promising type of digital counting is that using qualitatively different entities. In the process of embryonic induction, a tissue of kind A causes a second tissue with which it comes into contact to differentiate into kind B. This is exactly comparable to a 'rule of counting', whereby each number determines the characteristics of the next number.

Experiments suggesting a digital counting mechanism were performed by Moment [4] on the worm Clymenella, which has a constant number of 22 segments. If a few segments are removed simultaneously from the anterior and posterior end of the worm, it can regenerate back to the original number of 22. More relevant, the isolated piece retains its position in the series; thus if for example four segments are removed anteriorly and two posteriorly, exactly those numbers are regenerated.

These observations are inexplicable unless there is some difference between the segments. It is interesting though that Clymenella can count backwards as well as forwards. The simplest rule for this would be: 'If a C is in contact with a B, it causes the next segment to be a D; if a C is in contact with a D, it causes the next segment to be a B.' This brings us very close to Wolpert's problem of the French Flag.

References

1. A.M.Turing *Phil. Trans. Roy. Soc. B 237* (1952) 37.

2. J.Maynard Smith in *Mathematics and Computer Science in Biology and Medicine.* (HMSO 1965).

3. J.Maynard Smith *Proc. Roy. Soc.*, B *152* (1960) 397.

4. B.G.Moment J. *Exp. Zool. 117* (1951) 1.

The French Flag Problem: A Contribution to the Discussion on Pattern Development and Regulation

Lewis Wolpert
Middlesex Hospital Medical School

In trying to fit the process of development into a general theoretical framework it is convenient to distinguish between those processes having an essentially temporal quality and those having an essentially spatial quality. This distinction also reflects the way in which the problem is currently being approached. Most current work in development is directed towards the temporal aspect as manifested by differentiation of cells. More specifically it deals with the sequences of events involved in the control of the synthesis of specific macromolecules and is amply covered by other members of this meeting [1]. The spatial aspects of development, to which very much less attention is given, include regionalization or pattern formation, and morphogenesis.

A central problem in development is how the apparently structurally simple and undifferentiated egg can reliably give rise to an organism whose form is complex. In an earlier paper on the relevance of information theory to development [2] we suggested that the development of this complexity was more easily understood if the egg was considered to contain the programme for making the organism rather than the complete specification of the organism to which it would give rise. We suggested that it is much simpler to specify how to form a complex organism than to specify the organism itself. Some experimental justification for this view comes from studies on the cellular basis of morphogenesis in the sea urchin embryo [3] in which we suggested that much of the early changes in shape of the embryo can be accounted for in terms of the repeated occurrence of two cellular activities involving cell contact and pseudopod formation. It was clear that the specification required for these activities was much simpler than that which would be required for the shapes to which they gave rise. Moreover these cellular activities could provide the link between gene action and morphology.

As Waddington has put it, the problem in pattern formation 'is to identify the immediate causes of the breakage of a uniform region into separate elements situated in a definite spatial order'. While some insight into morphogenesis can probably be found along the lines just indicated, pattern formation is the least understood aspect of development [4]. What we require to know is the nature of the cellular activities involved in pattern formation, not only because it is an important developmental problem, but because it may be of significance in our whole conceptualisation of the nature of development. This latter point is particularly relevant to developmental regulation which is largely concerned with pattern formation. For, when one considers canalization in development, or regulation following removal of a part, it is the restoration of the spatial pattern to which one usually refers, and to which Waddington [5] refers to as homeorhesis On the other hand, current concepts of regulation such as end product inhibition and repression, are concerned solely with temporal aspects. (It should, however, be noted that both pattern formation and morphogenesis will always involve temporal differentiation.)

Our approach to the problem of pattern formation [4] has been to concentrate on simple axial patterns, both because they are simpler to analyse and because much experimental work has been carried out on systems where the pattern is essentially axial. The nature of the problem is best illustrated by some specific examples. (A) In sea urchin development, the early embryo becomes divided along the animal-vegetal axis into ectoderm, endoderm and mesenchyme. While the relative proportions of these regions can be varied both by operative and chemical means, the system has a marked tendency to regulate so that the normal proportions are maintained. For example, if those cells of the embryo, the micromeres, which normally gives rise to the primary mesenchyme, are removed, the embryo regulates so that the adjacent region gives rise to a more or less normal number of primary mesenchyme cells [6]. (B) The cellular slime mould can develop from a single amoeba liberated from a spore. This amoeba can, by growth and division, form a population which eventually aggregates to form a cartridge-shaped mass – the grex – which, in *Dictyostelium*, migrates. The anterior cells become stalk cells and the posterior cells spores, and the proportions of these two types may be invariant over quite a wide range in cell number. Moreover, removal of portions of the grex can leave the final proportions of stalk cells to spore cells unaffected [7]. (C) Hydroids have, in many cases, an essentially axial organization; for example, *Tubularia* is divided along its axis into oral cone, distal tentacles, gonaphores, proximal tentacles, and stolon. Hydroids show considerable powers of regulation in that they are able to reconstitute this pattern when a portion is removed or isolated [8, 9].

The capacity for the regulation of an axial pattern as illustrated by these examples can be observed in many other systems. Some of the most characteristic features are: (a) the pattern as manifested by the proportions between the parts is, within limits, invariant with size; (b) part of the system can produce the whole, and again within limits, any part can become any other part; (c) the polarity of the system is maintained, that is the parts of the pattern are formed in the correct positions relative to each other and to the pattern in the original system. We have formalized these aspects of pattern regulation so as to make it more amenable to theoretical analysis. The behaviour of such systems should be contrasted with the so-called 'mosaic' or non-regulative pattern formation in which removal of a part leads to a defect in the system, which is strictly related to the part removed. However, as Weiss [10] has emphasized, the distinction between regulative and mosaic patterns may only reflect the time at which regulation no longer takes place. Mosaic patterns are usually interpreted in terms of a prepattern, and one may conjecture that one would always find a regulative phase.

The experimental approach to this problem has led to the formulation of an interesting and simplifying set of concepts involving gradients and dominance [11, 9]. However, it is striking how little attention has been given to the theoretical side of the problem, particularly to the construction of working models. I know of only one serious attempt to construct a model which will develop and regulate an axial pattern, that put forward by Rose [12]. Within a more general theoretical framework, Spiegelman [13] is one of the very few who have given some attention to this type of problem. He pointed out that one of the features of regeneration is that the potentiality for forming a particular part of the pattern is present in a larger area than actually forms the part. Spiegelman suggested that such a system requires 'a principle of limited realization'. He considered that this must be brought about by at least two distinct mechanisms, the first involving suppression of realization of potentialities, and the second providing for differences between the parts of the system. He argued that if, for example, two parts of the system are capable of forming a particular region, and only one does, then it is essential to have both a difference between these two parts, and the suppression of the development of the one part by the other. He stressed that a difference between the two parts, as manifested by, for example, a gradient of some type, does not by itself provide an adequate mechanism. For some time we thought these conclusions to be correct and regarded them as constituting a First Theorem in pattern formation. However, from our analysis it has become clear that it requires considerable modification; the principle of limited realization only applies to the spontaneous self-limiting reaction (SSR) and there need not be continuous differences between the parts of the system.

▸ *Formalization of the problem*. Consider an axial array of N similar units. Each unit is capable of exhibiting a number *k*, of mutually exclusive states S_p ($p=1 \ldots k$). The state S_p of any unit is determined uniquely by the inputs to the unit and its previous state. The problem is to determine the inputs, outputs, and properties of the unit, and the nature of the communication between them such that the system becomes subdivided along its length into r regions, in a fixed order, each containing n elements, and characterized by a specific state S_p, such that $n_p/N=A_p$ (i.e. the number of units in a given state is a constant fraction of the total number of units). This configuration must be maintained however the system is disturbed by removal of parts. We have found it convenient to consider the case for $r=3$, and for $A_1=A_2=A_3=1/3$. If S_1, S_2, S_3 are blue, white, and red respectively, which will be denoted by S_b, S_w, and S_r, then the pattern is that of the French Flag, and this pattern must be restored whichever part is removed (fig. 1).

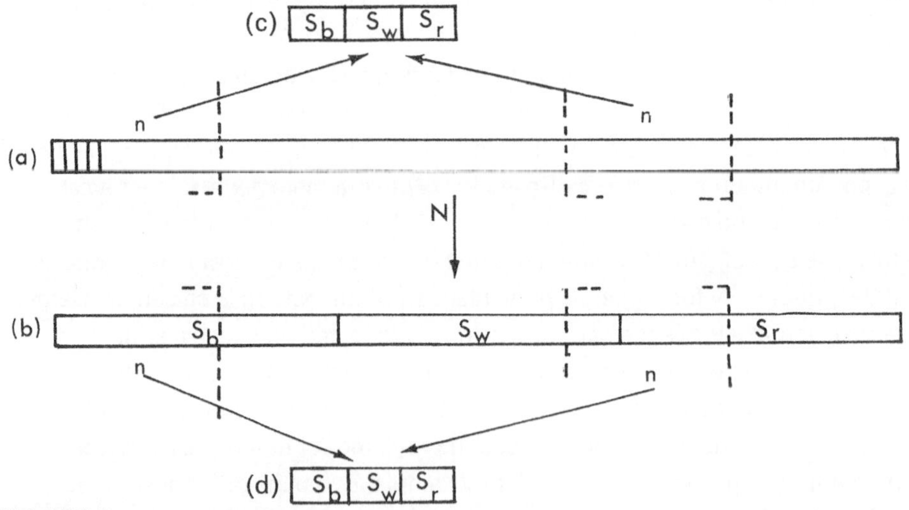

FIGURE 1
In (a) N similar units are arranged in an axial system – though only a few of the units are actually drawn in at the left-hand end. Each unit can be in state S_b or S_w or S_r. The system develops into the French Flag in (b) since the left-hand third is blue, the middle third white, and the right-hand third red. If a portion of the system in (a) comprising n units is isolated, it will develop into the French Flag as in (c). If part of the system which has already formed the French Flag is isolated, it will regulate to form a smaller flag as in (d)

This problem may be regarded as the paradigm for pattern formation, and it can be seen that it is an idealization of the biological examples given above. (For a more difficult problem consider the two-dimensional case – the Union Jack.)

There is undoubtedly some general method by which a solution to this problem might be found, and I very much hope that someone will suggest one. Since at present we do not have such methods at our disposal, our approach has had to be much more pedestrian. What we have done is to try to construct simple models that work and see what sort of general properties they have.

▸ *Model systems.* Before considering our own models, I should refer to the mechanisms suggested by Rose [12], not only because his is the only serious attempt at model making but because the ideas in his model are repeatedly implied, either explicitly or implicitly, when pattern and gradients are discussed. Rose suggested that the genesis of a pattern could be a consequence of a hierarchy of self-limiting reactions, and the spread of restrictive or inhibitory information from one differentiating region to another, together with a gradient in rate of differentiation, equivalent to Child's axial gradients. In this system, the reactions will proceed fastest at the high point in the gradient and the reaction which will predominate is that at the top of the hierarchy. This reaction is self-limiting and so allows the next reaction to proceed in the adjacent region, and so the gradient in rate provides a means for the spatial expression of the hierarchy in reactions. While at first sight this may appear attractive, an analysis of the system shows that it is a very unsatisfactory system for generating pattern and is incapable of regulation [4]. More generally, we have been unable to develop any model which relies on a gradient in rates of reaction. An even more important feature of Rose's model is that the size of a region is determined by the length of time for which the reaction proceeds; in our hands this has proved an unworkable basis for the French Flag.

From our attempts at the French Flag problem, there seems now to be two main ways for solving it. In the first type there is a mechanism whereby a unit can compute exactly where it is in the system and can thus 'decide' what state it should be in. A simple mechanism whereby this can be provided is by having two accurate gradients generated from the two ends of the system. The unit can compute its position from the value of the two gradients and so enter into the appropriate state. This type of model will be called the 'fixed gradient' type. In the second type of model the unit cannot compute where it is in the system in relation to the two ends. The mechanism relies on 'voting' or 'balancing' of the sizes of the regions against each other. This is seen in Webster's [14] threshold gradient model. In this, one region S_b produces a freely diffusible substance and the adjacent region S_w destroys it. Units change from state $S_b \rightarrow S_w$ reversibly until the concentration of the substance remains constant. In Webster's model there is a threshold gradient which provides that the units that change state are at the border between region S_w and region S_b. The proportions of the regions are determined by the rates of production and destruction of the diffusible substances. Two models will now be considered in somewhat more detail.

Before doing so it is useful to point out the distinction which may be made between global or system properties which belong to the system as a whole or which are only possessed by the individual units when in the system; and local or unit properties. One should also be cautious in too readily identifying units with cells.

▶ *A fixed gradient model.* The units are assumed to have fixed thresholds T to a single substance I, such that the concentration determines the state of the cell. Thus, if $I > T_b$, the unit will be in state S_b; if $T_b > I > T_w$, then the unit will be in state S_w; if $T_w > I$, then the unit will be in S_r. These are the local properties of the system. The global properties involve the establishment of a linear gradient whose absolute values at either end is regulated and fixed at I_l and I_r respectively (fig. 2). Then if

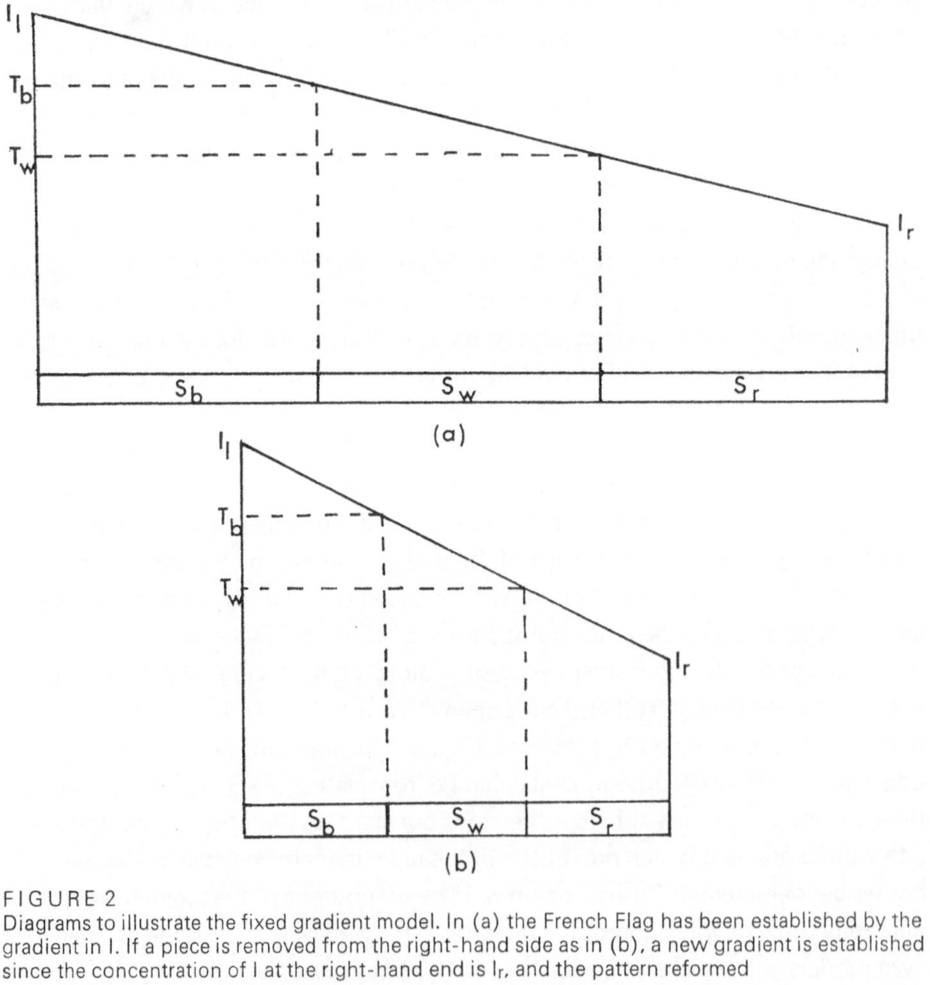

FIGURE 2
Diagrams to illustrate the fixed gradient model. In (a) the French Flag has been established by the gradient in I. If a piece is removed from the right-hand side as in (b), a new gradient is established since the concentration of I at the right-hand end is I_r, and the pattern reformed

$I - T_b = T_b - T_w = T_w - I_r$, the system will form the French Flag. The system will regulate perfectly if, when a piece is removed, the values of I at the ends are restored to I_l and I_r and the gradient in I between them is linear (fig. 2). This regulation can be achieved if the unit at the left-hand end is the sole source of production of I and the rate of production is controlled so that its value there is I_l, the unit at the right-hand end is a sink for I, and the rate of removal is controlled such that the concentration of I there is kept at I_r. Under these conditions the substance I will diffuse from the left-hand to the right-hand end and the concentration gradient will be linear [4]. It is clear that this system requires a mechanism whereby a unit's behaviour depends on whether its left-hand end is free (source), neither end is free, its right-hand end is free (sink). This of course implies that each unit is polarized.

This system, which is capable of perfect regulation, has several interesting characteristics. The threshold properties of the units are fixed and only one substance is involved, irrespective of into how many regions the system is divided up. Particularly interesting is the fact that there is no interaction between the regions S_b, S_w, and S_r and the size of the one does not affect the size of any other. Intercellular communication is limited to the units having input from their neighbours: if they have a unit on either side, they are inhibited from either producing or destroying I; if they have a unit only on their right side, the destruction of I is inhibited and I is produced so that its concentration is I_l; if they have a unit only on their left side, the production of I is inhibited.

▸ *A balancing model.* This model has features similar to Webster's [14] threshold gradient model but does not require any gradient. Consider first the subdivision of the system into two regions S_b and S_w such that $A_b = 1/3$ and $A_w = 2/3$. The units spontaneously enter into S_b and produce a substance I_b at a rate k_b/unit and this induces cells to enter S_w and produce I_w at a rate k_w. Both I_b and I_w are rapidly diffusible and have similar short half-lives. The rule for entering S_b or S_w is that if there is an excess of I_w over I_b, units will enter S_b; if I_b is in excess of I_w, then units will enter S_w; but only a unit which is adjacent to units in both S_b and S_w, that is at the border, can change its state. The only exception to the latter rule is that a unit at the left-hand end can enter S_b from S_w if $I_b = 0$ and a unit at the right-hand end can enter S_w from S_b if $I_w = 0$. Clearly units at the border will change state until the concentration of $I_w = I_b$, so if $k_b = 2k_w$, then the system will regulate such that $A_b = 1/3$, $A_w = 2/3$. This can easily be extended to subdivide the units in S_w so as to form the French Flag, which will regulate perfectly. This type of model may be contrasted with the previous one. Not only can a unit never compute where it is in the system, but the system has no gradient. This latter point is of great importance, as it shows that gradients are not required for pattern regulation.

▸ *General principles*. These two types of models seem to provide quite reasonable solutions to the French Flag problem, and it is possible by looking more closely at some of the features, to find some features which are common to all the models that solve the French Flag problem. There seem to be three necessary conditions. Firstly, it is clear that the system must be polarized, that is every unit must have a mechanism whereby that end of the system which is in state S_b is specified: to make the Flag, the end which is to be blue must always be specified. The polarity of the system may require the units themselves to be polarized, but polarity may be a global property. The second feature common to all models is the necessity of the units having threshold properties. This means that the state of the unit will depend on certain values of the inputs, and will switch from one state to another at critical values. These thresholds may depend not only on the actual inputs but on some complex function dependent upon them. The thresholds may be unit properties as in the fixed gradient models or global, as in the gradient threshold model of Webster [14]. The third condition, common to all the models, is less easily defined. I would suggest that it is the requirement for at least one spontaneous self-limiting reaction (S S R). By spontaneous one means that the reaction will proceed in any unit unless inhibited from doing so by some other part of the system; the self-limiting aspect refers to the fact that the inhibition arises from the reaction having proceeded elsewhere in the system, and this provides the inhibitory mechanism. In the fixed gradient model there are two S S R's – the production of I and the removal of I. The reactions are localized at the ends of the system. In the 'balancing' model, the S S R is to enter state S_b. This is self-limiting, since units in S_b produce I_b, which can cause S_b units to enter S_w. The S S R is the only part of the system that requires Spiegelman's principle of limited realization, but does not require graded differences between the parts. It should also be noted that the S S R provides the basis for dynamic control.

While I am extremely conscious of the naivety of my approach to the French Flag problem, it seems so far that three conditions are necessary (and perhaps sufficient) for its solution, namely polarity, thresholds, and S S R. I very much look forward to someone providing a much more general and sophisticated approach. It will be important to establish whether these three conditions are in fact necessary requirements for the development and regulation of the French Flag. If they are, then they provide a basis by which models may be judged and, more important, we can try to identify them in biological systems; it is tempting, for example, to try to equate the SSR with the concept of dominance.

It is of interest to consider the regulation of the French Flag in relation to homeostasis and homeorhesis. Since we have treated, by definition, the French Flag as a terminal pattern, which does not change further, we have made it inevitable that its formation should be considered as an example of homeostasis rather than of homeorhesis, since the system always returns to a constant state rather than a developmental pathway. (In fact it is not clear to me just what a developmental pathway is in cellular terms.) In practice, many French Flag patterns in biology are not simply terminal. Thus of the three examples mentioned in the introduction, obviously the early echiniderm embryo and the slime mould grex are transitory stages in more extended developmental sequences. It is thus necessary to ask to what extent development of pattern involves a succession of French Flags whose regions develop into Union Jacks, and so on. It would require still further theoretical development to treat such problems of longer-term pattern formation and modification. However, it is hoped that the discussion we have given of a simpler system which involves only structural homeostasis may be a useful basis from which to proceed. In conclusion, it may be remarked that notions such as homeostasis, homeorhesis, and the epigenetic landscape can be nothing more than descriptions of the logical properties of systems, and do not attempt to provide any insight into the mechanisms by which those properties are achieved. They can therefore make no reference to the factors of polarity and SSR which I believe to be the special properties of pattern regulation, and the spatial aspects of development.

This work is supported by the Nuffield Foundation.

References

1. B.C.Goodwin *Temporal Organization in Cells.* (Academic Press 1963).
2. M.Apter and L.Wolpert *J. Theoret Biol., 8* (1965) 244.
3. T.Gustafson and L.Wolpert *Int. Rev. Cytol., 10,* (1963) 163.
4. L.Wolpert and G.Webster In preparation 1968.
5. C.H.Waddington *The Strategy of the Genes* (Allen & Unwin 1957).
6. S.Hörstadius *Biol. Rev.* (*Cambridge*) *14* (1939) 132.
7. K.B.Raper *J. Elisha Mitch Sci. Soc. 56* (1940) 241.
8. G.Webster and L.Wolpert *J. embryol. exp. Morph. 16* (1966) 91.
9. J.S.Huxley and G.R.de Beer *The Elements of Experimental Embryology* (Cambridge University Press, 1934).
10. P.Weiss *Principles of Development* (Holt 1939).
11. C.M.Child *Patterns and Problems of Development.* (University of Chicago Press, 1941).
12. S.M.Rose *Amer. Nat. 86* (1952) 337.
13. S.Spiegelman *Q. Rev. Biol. 20* (1945) 121.
14. G.Webster *J. embryol. exp. Morph. 16* (1966) 123.

The Division of Cells and the Fusion of Ideas

Brian Goodwin
University of Sussex

The two main types of computer currently available, analogue and digital, reflect with remarkable fidelity two quite different approaches to the analysis of biological systems. The analogue approach emphasizes the dynamic, largely continuous aspects of biological process such as physiological activity and ecological balance, while the digital approach emphasizes the quasi-static, discontinuous, logical aspects of biological phenomena such as speciation in genetics and decision-making in psychology. Fields of study which constantly run up against both types of behaviour, such as embryonic development and learning, are particularly intractable from a theoretical point of view, and any complete treatment of such phenomena must obviously incorporate both sides of their behaviour, continuous and discontinuous.

In the world of computers a marriage between the two types of machine is not difficult: you make a hybrid. It is perhaps suggestive that in the hybrid the two components are not equal partners; usually the analogue part is made the 'slave' of the digital component, which functions as the 'brains' of the machine. The relationship is somewhat as the id to the ego. Which partner is regarded as ultimately ruling the whole machine depends upon the point of view, and whether your sympathies lie with Bergson or with Russell.

In the world of mathematics, a marriage between automata theory and control theory, which roughly speaking define the different emphases of approach, is not so easy to arrange. This is a major task, and on its resolution, it seems to me, hangs much of the future of theoretical biology. I would like to consider rather briefly and somewhat superficially a particular biological phenomenon, the cycle of cell growth and division, in order to illustrate the two complementary aspects of the process and how they may be handled analytically by automata theory and by control theory. I will also suggest how these may be unified, and what problems are encountered in so doing.

▸ *Discontinuities in biological systems.* In the areas of biology where biochemistry is obviously important and concentrations of substances play a role in determining the behaviour of the system, discontinuities are most commonly explained as arising from the existence of a threshold. Thus, for example, the differentiation of a simple organism such as *Hydra* into two distinct regions, head and stalk, is usually explained by assuming that there is a unidirectional, continuous concentration gradient of some substance running along the axis of symmetry of the organism, and a threshold 'detector' in each cell. All cells experiencing a concentration above threshold differentiate into head cells, while these experiencing sub-threshold concentrations differentiate into stalk cells. In fact, this model is too simple to fit the facts of differentiation and regulation in *Hydra*, but it serves to describe the way biologists think about threshold phenomena in concentration-dependent systems. This type of threshold response to continuously varying concentrations is also used to explain 'decisions' taken by cells in the course of the cell cycle, such as when to initiate a new wave of DNA replication. (In this case the concentration change is in time rather than in space.) Thus the logical features of spatial and temporal decisions taken by cells are regarded as arising from biochemical switching mechanisms.

Exactly how such switches work at the molecular level is not entirely clear, but it would seem that co-operative behaviour of multimeric macromolecules, most commonly proteins, is the most likely current candidate. At a critical concentration of protein monomers, a rather sudden aggregation of protein can occur producing dimers, trimers, tetramers, or higher-order multimers. This process is like a phase transition in a physical system, since the multimers can have quite different biochemical properties from the monomers. Because small molecules and ions can affect the equilibrium position of the aggregation-disaggregation process, gradients or temporal concentration changes of such molecules can determine a threshold point for aggregation quite sharply. Thus a more or less continuous biochemical substratum of small molecules, operating on macromolecules and macromolecular structures, can produce the spatial and temporal heterogeneity which allows one to look upon the cell as an automaton or an ensemble of molecular automata which act entirely on the basis of binary decisions. One can then represent the cell as a type of Turing machine in the manner of Stahl [1].

This procedure allows one to focus attention on the logical features of cell behaviour. For example, one can ask if the cell as represented in automaton form is capable of Lamarckian-type evolution, wherein an adaptive structure or molecule must be encoded into the genome and concomitantly calculated to be consistent with all extant cellular properties. Such a calculation runs into certain unsolvable problems, according to Stahl [2]. Thus it may be possible to demonstrate that Lamarckian evolution cannot occur in an automaton modelled on the adaptive and self-replicative properties of cells – a result of some interest. It is also suggested that cancer may be explained as the inability of a host organism to identify a malignant cell as abnormal, again because of an unsolvability situation. Whether or not this approach can shed new light on this problem of cancer, for which there are many less sophisticated explanations, can only be resolved by further evidence.

▸ *Dynamic aspects of cellular processes*. An automaton model of cell growth and division has dynamic features, but they are of a highly discontinuous, stepwise nature. The sequence of events making up the cell cycle as described by Stahl is of a simple causal type, represented by decisions which result in one gene after another being read, then the genes being shut off, the DNA replicated and partitioned between daughter cells, etc. This is a perfectly legitimate idealization of cell function for purposes of certain logical and computational questions. However, many of the main problems regarding the organization of molecular processes underlying the cell division cycle concern the details of gene interaction, the timing of gene read-out, and variations in these resulting from different environmental conditions. Automata theory is not primarily concerned with questions of this kind, and starts from a somewhat different level of analysis. Nevertheless some of these problems must be resolved before an adequate theoretical treatment of the cell cycle can be given. The assumption of sequential gene read-out, for example, is one requiring direct experimental investigation, for it appears that some genes are transcribed continuously while others are transcribed discontinuously. Since as many as 2,000 different genes may be read in the process of making two bacteria from one, which requires about 50 minutes (minimal medium with glucose), it is clear from the rates of macro-molecular synthesis (about one second to make one molecule of mRNA or protein) that all genes cannot be read out in strict sequence. Genes must be read either in overlapping sequence or in groups so that many are transcribing simultaneously.

A fundamental question in relation to control dynamics in cells is whether or not gene action has all-or-none character, a gene being full on when its repressor level is lower than a particular threshold value and shut off when it is above this value. Such on-off characteristics would make the gene a two-state module and the cell an essentially digital-type mechanism at this level of control. However, the evidence available is fairly conclusive in demonstrating that those genes so far investigated (for alkaline phosphatase and β-galactosidase) can be continuously regulated over an order of magnitude by continuous variation of repressor level. This transforms the problem of gene regulation into one of control rather than strictly logic, although at higher organizational levels threshold devices are evidently operative (e.g. the embryological switch regulating differentiation and morphogenesis in *Hydra* mentioned earlier).

Since gene activity is neither strictly sequential nor on-off, the temporal ordering of some genes or sets of genes during the cell cycle is transformed into a problem of phase relationships between oscillators rather than the sequential operation of on-off switches. However, it is necessary that the dynamic representation of gene activities include in it the possibility of highly discontinuous behaviour, such as that characteristic of a relaxation oscillator. Evidence from the behaviour of biological clocks in protozoa such as *Euglena* or *Gonyaulax* (wherein the clock phenomenon and the cell cycle are intimately connected) points very strongly to this type of highly non-linear behaviour.

Thus in seeking a molecular model for the processes underlying the cycle of growth and division in cells one is forced into the middle ground between the continuous, linear realm which has been so useful and fruitful in analysing physical systems where additivity is the rule; and the totally discontinuous, highly non-linear realm of Boolean logic and automata theory which has flourished with the advent of digital computers. Early in this century physics seemed about to be forced into this difficult mathematical terrain, with the anomalous properties of black-body radiation, photoelectric effects, and atomic spectra. But the discovery of the universal fine-structure unit provided by Planck's constant meant that quantum mechanics could be developed by means of a quasi-linearization procedure, and the linear domain was extended to cover quantum phenomena. It is too early to say whether or not such a procedure may be possible in biology, for example in relation to the dynamics of the control processes operating during the cell cycle. It may be that rules will be found which relate some function of frequencies and amplitudes of periodically varying quantities such as enzyme activities, in relatively simple ways, with exclusion principles and forbidden phase relationships defining the ordering of events during the cell cycle. Some primitive principles of this kind are already emerging from work I am doing on phase, frequency, and amplitude relations of several enzymes in synchronized continuous cultures of *E. coli*. Many of the 'rules' observed are simply those governing the interaction of non-linear oscillators, but they are expressed in a semi-empirical manner. Instead of trying to deduce the detailed behaviour of cells from differential equations representing molecular control processes, a more heuristic procedure is being followed which is already a step in the direction of an automaton description. The latter emphasizes logical sequence of events but ignores energy interactions. The only 'variables' it uses, then, are frequencies and phases of events, ignoring amplitudes or interactions, and even these two variables are rigidly fixed. An intermediate position between that of an excessively detailed differential-equation description and an automaton treatment seems to me the most likely to be successful in dealing with this problem.

Towards a resolution. One of the pitfalls of a continuous dynamic analysis of cell behaviour is that one can get thoroughly lost in the bewildering detail of variation in system variables and fail to abstract sufficiently to catch only the relevant features of their behaviour. An automaton description of molecular organization starts from a highly abstracted representation and so avoids this difficulty; but, as I have already observed, it imposes a rigidity of behaviour on the system which may fail to capture the basis of its organization. The trick is to select the features of the variables and their interactions which preserve their essential properties while ignoring irrelevant details. The energy function of physics serves this purpose. I feel that biological processes require for their description and analysis an analogous function, but one relating to units of control activity. The talandic energy function which I described in a strictly theoretical context a few years ago [3] was intended to serve in this capacity. I have found this kind of function useful in expressing the amount of control and informational activity in a regulation unit and its influence on other units in the cell. This influence or interaction involves both frequency and amplitude modulation.

This procedure will lead, I hope, to the definition of a state function whose minimum defines the 'configuration' of the system, specified ultimately in terms of frequencies, phases, amplitudes, and other essential characteristics of cell variables required to describe the dynamic state of a growing and dividing cell. These quantities will then define the temporally ordered, periodic set of events which constitute the self-replicating morphogenetic process known as the cell division cycle. The sequential nature of the biosynthetic and morphogenetic events, emphasized by an automaton description of the process and regarded by Pattee [4] as essential to a self-replicating cycle, is clearly a major feature of such a description. In fact the logical aspects of the system's organization would all be present, but supplemented by additional properties relating to the more continuous, regulatory features of the system. So far relatively few details of this partially abstracted, non-linear dynamic description of the cell cycle have been worked out. Experimental results will soon show if semi-empirical procedures are successful in simplifying the non-linear aspects of control dynamics and allow the use of quasi-linear operations in the description and analysis of their behaviour. This approach does seem to provide the possibility of including the essential features of the analogue and the digital properties of the biological system under investigation in a coherent theory. How useful it will be in other areas such as embyonic development and learning remains to be seen.

References

1. W. R. Stahl *J. Theoret. Biol. 8* (1965) 371.
2. W. R. Stahl in *Natural Automata and Useful Simulations* (Macmillan 1966) pp. 43–72.
3. B. C. Goodwin *Temporal Organizations in Cells* (Academic Press 1963)
4. H. H. Pattee in *Natural Automata and Useful Simulations* (Macmillan 1966) pp. 73–105.

Tolerance Spaces and the Brain

E. C. Zeeman and O. P. Buneman
University of Warwick

Introduction. Tolerance spaces are a mathematical tool useful for describing the brain. We first show how they arise in making a model of the brain, and then at the end of the paper give some precise definitions and properties.

A characteristic feature of any brain is to make internal models of the outside world. A stimulus x from the outside world causes the brain to get into an internal state y, say. Therefore if X denotes a set of external stimuli, and Y denotes the corresponding set of states of the brain, we have a map (or function)

$$f: X \rightarrow Y$$

Of course the stimulus x may be totally unlike the corresponding state y; for example, x might be a picture and y a 'wave pattern' in the cortex. Nevertheless if the brain is to make any sense of all the incoming messages, the way in which the different stimuli are organized and related to one another must be echoed to some extent in the way in which the corresponding brain states are related to one another.

In mathematical language the sets X, Y must have some structure and the map f must preserve this structure. Two questions arise immediately:

1. What sort of structure?
2. What are 'states' of the brain?

The set of stimuli are easier to understand – we should like the theory to cover such things as visual images, tactile images, sounds, colours, or smells for example.

▸ *Mathematical structure.* There are many types of mathematical structure, for example symmetry in visual patterns or rhythm in music, but the structure that we want to examine is a much weaker and more general kind of structure, which we shall call a *tolerance*. It corresponds to the psychologists' 'least noticeable difference'. Suppose X is a set of stimuli. If two stimuli x_1, x_2 in X are sufficiently close for us not to be able to distinguish between them we say they are *within tolerance*, and we write $x_1 \sim x_2$. Conversely, if we can distinguish between them we say they are outside the tolerance. The *tolerance* ξ is defined to be the set of pairs (x_1, x_2) such that $x_1 \sim x_2$. The set X together with the tolerance ξ is called a *tolerance space*.

For example, if X is the visual field, which is roughly a quarter of a sphere, then two points near the centre of the visual field are indistinguishable, or within tolerance, if they are less than 1 minute of arc apart, while towards the periphery visual acuity lapses, and so the angular distance increases to about 1 degree.

The intuitive idea of doing mathematics on a tolerance space is that we can move about within the tolerance without noticing any difference; when we have to prove two things equal it is only necessary to do so within tolerance, or if we wish to investigate the structural stability of a differential equation it is only possible to do so within tolerance, and so on. At first sight it looks like a serious handicap, but after playing with the idea for a while it reveals considerable flexibility. Another intriguing aspect is that the tolerance has the effect of gluing X together, in much the same way that a topology glues a topological space together. Perhaps the notion of gluing points together is a slightly misleading analogy because a tolerance is not in general transitive, that is to say if $x_1 \sim x_2$ and $x_2 \sim x_3$ this does not imply that $x_1 \sim x_3$. It is very important that a tolerance should not be transitive because as we near the threshold of the least noticeable difference we always run into non-transitivity.

On the other hand, sometimes the concept of a set X being glued together by the tolerance is intuitively valuable. For example, if X is the set of colours (which from the point of view of physics is an ∞-dimensional function space) and ξ is the tolerance given by human colour perception, then ξ has the effect of gluing X together into the familiar two-dimensional colour chart, together with a factor for intensity.

Many of the things we can do with topological spaces we can also do with tolerance spaces, and for this reason we borrow both words and notation from topology. In particular we can use the strong global invariants of topology and at the same time neglect local properties, because the latter are all within tolerance. Therefore a tolerance space is like a blurred topological space. And this is exactly the kind of mathematics that ought to be useful in studying the brain, because any model of the brain ought to be based firmly on the detailed neurological structure; and yet, having constructed the model, we want mainly to investigate the global properties in order to study thinking and memory, at the same time ignoring the local details to within tolerance.

▸ *Dynamic states of the brain.* There are two distinct things one can mean by a state of the brain at a given time. The first is the *dynamic* state, which involves the chemical balance and the rates of firing of all the neurones at that time. Experiments indicate that the dynamic state is related to what is being thought or felt, and so we may crudely assume the dynamic state determines the thought.

The second is the *static* state, which involves the network of the 10^{14} neurons and the strengths of the 10^{15} synapses plus maybe many other things. We should expect that memory is stored in the static state (potential memory, that is, because realized memory is involved in the dynamic state). For example, during sleep the dynamic state might be dominated by the sleep rhythm, but the static state would still store all memory.

Let Y denote the set of dynamic states of the brain. We should expect that sufficiently close states y_1, y_2 would determine the same thought, and so, writing $y_1 \sim y_2$, we must have an intrinsic tolerance η on Y. The dynamic states therefore form a tolerance space (Y, η). The reader will protest that we have not yet defined what exactly a dynamic state of the brain is – but we claim this does not matter. For if we had two tolerance spaces, (Y, η) and (Y', η'), both purporting to describe the dynamic states of the brain, then there would be a tolerance homeomorphism (see definition at the end of the paper)

$$g : Y \to Y'$$

between them. In other words, the two representations would be identical, to within tolerance. Therefore the notion of tolerance space enables us to choose with impunity the most convenient form of representation – with one proviso that it be sufficiently detailed.

The particular Y suggested was a unit cube of dimension 10^{14}. Each point $y \in Y$ has co-ordinates (y_i) which represent the firing rates of the 10^{10} neurons, suitably parametrized. Thus the state is determined by the firing rates of all the neurons. The evidence that this set Y is sufficiently detailed is that the brain (and presumably the thought) is unaffected by the loss of a few thousand neurons. In other words, although the cube has sides of length 1, the tolerance is probably comparatively large – say any two points a distance less than 100 apart are within tolerance (the diagonal of the cube has length 10^5).

In particular this model ignores the activity of other parts of the brain, such as the glial cells, but we suggest that these can be ignored by virtue of the remarks above. It also ignores phase differences between the firing patterns of different neurons. But we also suggest that this does not matter within tolerance, because the effect of a burst of pulses in one neuron is primarily additive [1]. Where phase differences become crucial is among the slow waves engendered; experiments [2] suggest that phase differences around 10/sec. and below are important (as opposed to firing rates of 50–1000/sec. in any individual neurone). An analogous situation occurs in broadcasting, where we are not interested in the phase of the carrier radio wave, but we are interested in the phase of the carried sound wave.

Now the slow waves are visible in the model because as time t varies the point y(t) representing the state of the brain moves around the cube Y. Slow waves of the order of 10/sec. would be represented by periodic or 'helical' motion of y(t).

A 'helical' motion of y(t) might be periodic in some co-ordinates (representing perhaps a control group of neurons) and progressive in other co-ordinates (representing perhaps a sequential memory like a tune). Experimental evidence connects such neurological behaviour with memory both in cats and humans [3].

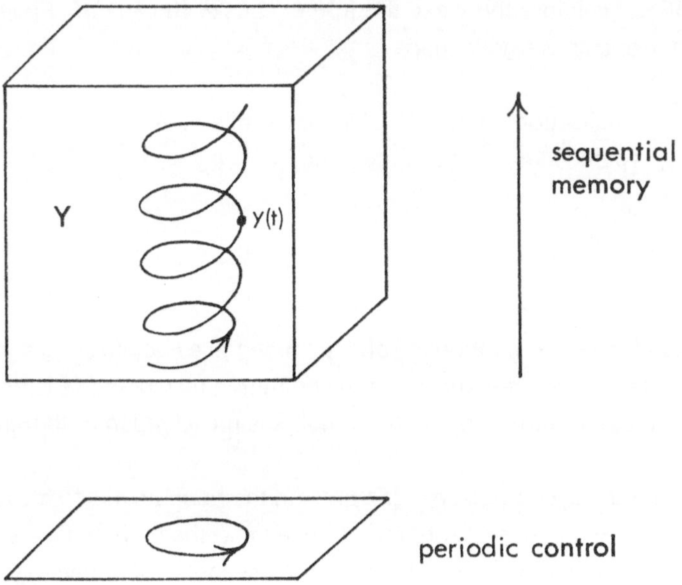

FIGURE 1

▸ *Memory fields*. What makes the point y(t) flow in a particular direction? The answer is that we have a vector field v on Y (in other words, an ordinary differential equation) and y(t) flows along the trajectories of this field, to within tolerance. The field v is determined by how all the neurons act on one another, and so into this one single symbol v we pack all the information about all the strengths of all the synapses, their threshold values, whether they are excitory or inhibitory, and how they act in concert under all possible circumstances. In other words, v is our potential memory.

Let V denote the set of vector fields on Y. As before, V has an intrinsic tolerance on it, but it is worth remarking that fields within tolerance may have trajectories that start within tolerance and quickly run out of tolerance, implying an element of unpredictability in our thinking and remembering.

The vector field v in V depends on time t and the input stimulus x. Therefore at time t the map $x \to v(x)$ gives a map

$$v : X \to V$$

This map is the potential memory of the stimuli, and tells us how the brain will react to any stimulus whatever it happened to be thinking previously. In other words, v represents the complete memory at time t.

Suppose now that the brain is in a prepared state of thought z in Y; then the vector field v will guide it into the 'next' thought y. Thus each vector field determines a thought y, and so we get a map

$$w_z : V \to Y$$

which represents a 'realization' of memory. We have called the map w_z because it depends upon prepared state z. The composition

$$X \xrightarrow{v} V \xrightarrow{w_z} Y, \qquad X \xrightarrow{f} Y$$

is the original map f that we first thought of. Experimental evidence suggests that both v and f are tolerance embeddings (see the definition below and [4]). In other words, the tolerance structure on the stimuli X is preserved in both thinking f and memory v.

Of course, when the psychologist experiments on the brain he has to measure the (thinking) map f. But f depends upon both time t and the prepared state of thought z (which also depends on t). Therefore the psychologist is severely limited to experiments that are as independent of z and t as possible. From his results on f he deduces properties of the (memory) map v, which is much more interesting. Indeed, not only is v a tolerance embedding, but v also attaches to each stimulus x a whole ambience of association, because the field v(x) provides the paths for the thought to flow along, and almost instantaneously recalls those associations.

The tolerance on V also explains how distant memories become blurred. As new memories are established, the fields representing the old memories are distorted and pushed within tolerance. Looking at it the other way round, the tolerance on the old memories increases, thereby blurring the details. So childhood memories become blurred with time. The concept of memories becoming blurred is much closer to reality than 'bits of memory-information' becoming lost.

▸ *Elaboration of the model.* The main idea is to take into account the different parts of the brain, and we shall show how the notion of tolerance spaces can facilitate

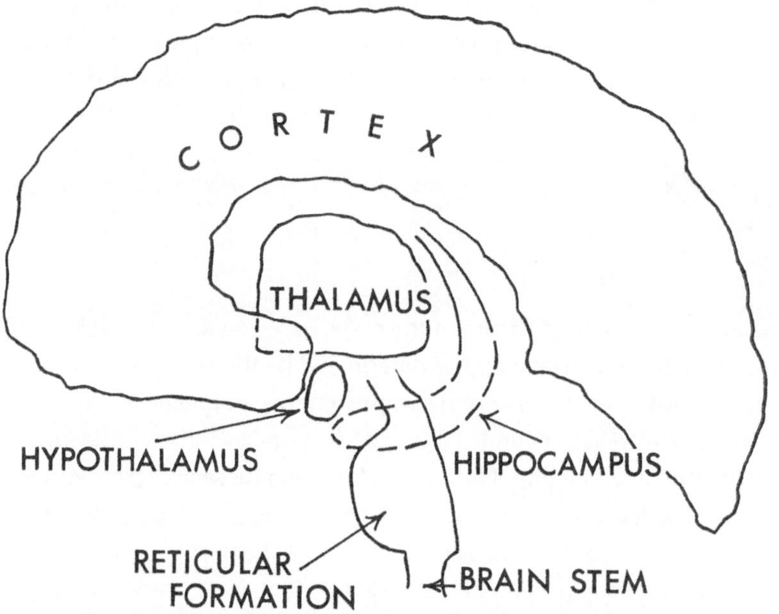

FIGURE 2

discussion. Our eventual aim is to produce testable hypotheses, though not in this paper.

The thalamus and other mid-brain regions differ from the cortex in that they contain a profusion of parasite synapses, whereas the cortex contains none. We use the term 'parasite' for the type of synapse described by Eccles [1] which is capable of altering the strength of the synapses on which it lies. We would speculate that these synapses are activated by other parts of the brain : the hypothalamus where the emotional centres lie and the so called 'pain' and 'pleasure' cells are found, or the arousal system of the reticular formation. The vector field v_T which describes the static state of the thalamic centres can therefore be easily and rapidly changed. On the other hand, there is no evidence to suggest that the strength of synapses in the cortex can be changed at all easily. Cragg [5], for example, finds that it takes a few hours of exposure to light to promote growth of synapses in the visual cortex of kittens when their eyes first open.

We use the terms 'soft' and 'hard' to denote these two types of synaptic arrangement. The sudden appearance of pain for example would immediately change v_T on the 'soft' thalamus without immediately affecting v_C on the 'hard' cortex. We would suppose that long-term memory involved a change of the 'hard' cortex.

▸ *Memory*. Anatomically we know that nerve fibres passing to the cortex pass exclusively through thalamic centres. We also know the thalamus is involved in all but the simplest reflexes. Let X denote the set of stimuli, and T and C the dynamic states of the cortex and thalamus. We have a diagram of tolerance maps

$$\begin{array}{ccccc} & & & & C \\ & & & {}^{g}\nearrow & \\ X & \xrightarrow{f} & T & & \downarrow k \\ & & & {}_{h}\nwarrow & \\ & & & & C \end{array}$$

Before we examine the possible nature of the maps g, h, and k, let us look at the implications of this diagram. The map k is determined by the static state of all cortical synapses; note that it does not distinguish between coding synapses (such as excite the Hubel Wiesel units [6]) and 'memory' synapses. Indeed, we would claim that the only functional difference between these synapses would lie in their relative hardness, and it is possible that a 'coding' synapse, by virtue of small variations in its strength, could also function as a memory synapse. The physiological or anatomical division of the cortex into 'analysers' and 'store' then seems rather artificial.

At first sight it looks as though X is the outside world and C is the inside world of our memory and that thinking takes place in T as it compares the two, but it is wrong to suppose that we have pushed the problem of thinking to the thalamus. We know that a simple response to say an auditory stimulus can take place independently of the cortex; a response to a visual stimulus would involve the cortex. In one case the thalamus responds to f, in the second to the composite map $f \cdot g \cdot k \cdot h$ (reading maps from left to right). Also we would claim that the thalamus could interact with the cortex more or less independently of the stimulus x, and it is this *interaction* that we tentatively label thinking.

What is the nature of the maps g and h? van Heerden [7] has pointed out an interesting analogy with laser photography, or the hologram.

In this picture P is transformed into a hologram H from which there is an inverse transform back to the picture

$$P \underset{h}{\overset{g}{\rightleftarrows}} H$$

The beauty of this method is that the storage in H is not local; removal of a part of H has no other effect than to blur P. Analogously the memory of particular stimuli does not seem to be specific to part of the cortex. Moreover, van Heerden's system is 'associative' in that if P has been stored then the composite map $g \cdot h$ of a picture P_0 which agrees with a part of P produces a 'ghost image' of the whole of P. Van Heerden's system requires that there is a coherent source which might correspond to a hippocampal rhythm such as that seen by Adey [2, 8]. It is uncertain that such a rhythm could give rise to the high spatial frequencies in the cortex that would be needed for good resolution of thalamic activity.

Going back to the brain, anatomically we know that the map g is caused by the cells that project from the thalamus to the cortex; the areas of such cells terminate in small regions of the cortex. Conversely, the projection of the cortex back to the thalamus is quite specific. This being so, the maps g and h in the absence of interference could be inverse to one another and possibly cause the α-rhythm. Of course, arousal or direction of attention constitutes interference, and thus the α-rhythm disappears.

There remains the problem of altering the hard vector field v_C of the cortex C. We have assumed that v_C is determined by the strengths of the cortical synapses and that these synapses can only modify as a result of the activity of the cells they connect. Hard synapses would presumably need maintained cortical activity to alter their strength. Gartside and Lippold [9] and others have seen that if the firing rate of a cortical neuron is increased artificially for a few minutes, its spontaneous activity remains at a high level for a considerable time.

This maintained activity would provide a method for transforming a dynamic state of the cortex into a new static state of its vector field. We could speculate that the method of producing this initial increase involves the hippocampus (see [2]), which might alter v_T in the thalamus in a way that causes v_T and v_C to 'resonate' until this increase is produced. We do not require a large change in each individual synapse; although each local change is small, the resulting global field has been changed outside the tolerance.

Abstractly the coupling of the hard cortical field v_C and the soft thalamic field v_T would give rise to a host of periodic phenomena. The form of these phenomena would alter with any change of v_T; thus it is observed that any change of the activity of the hypothalamus or reticular formation is immediately reflected in the EEG activity of the cortex. To clarify the nature of this coupling between the fields v_T and v_C we need to develop an understanding of generalized transforms and coupled differential equations in tolerance spaces. We need to understand how the coupling causes a decay in the resonance between the two fields at the same time as it bends the hard field (the transference from short term memory to long term memory); the situation is analogous to the transference of energy between coupled dynamical systems, although of course it is not 'energy' that is involved here. We also need to understand the continuous superposition of such changes in order to explain how continuous stimuli can be changed continuously into long term memory.

▶ *Language.* We emphasize the speculative nature of the preceding paragraphs. They are meant to illustrate how tolerance spaces can provide the foundations on which to build a very general theory of thinking and memory, globally rich enough to encompass the complete activity of a brain, and locally flexible enough to allow for almost any biochemical explanations. The only trouble with the model is that it is not yet specific enough to produce testable hypotheses, and so can be accused of being of little scientific value. Perhaps its main value is to provide a guideline on how to approach the problem from a global point of view.

Before leaving this speculative area, we show how tolerance spaces might be useful in discussing language. Ideas are sometimes distorted by putting them into words, and so this process imposes a tolerance on the set of ideas. Communication between two brains

ideas → words → ideas

must be inaccurate to within the product of the two tolerances. The further apart the two people are in upbringing, the larger the product of this tolerance, and the more inaccurate their communication. People who know each other reduce the tolerance considerably by adding overtones to the language by vocal and facial expression, and so increase the accuracy of communication. Different languages impose different tolerances, and the knowledge of two languages imposes a finer tolerance on ideas, thereby sharpening perception. Meanwhile translations by different people

ideas → French
English ↗↘
ideas → French

are inaccurate to within a tolerance at least twice as large as usual. Consequently one person's translation must necessarily be inaccurate to the other, however painstaking both translations are.

▶ *Tolerance spaces.* We conclude the paper with the definition examples and elementary properties of tolerance spaces.

Definitions. A *tolerance* ξ on a set X is a reflexive symmetric relation. Reflexive means that $x \sim x$ for all $x \in X$, and symmetric means that $x_1 \sim x_2$ implies $x_2 \sim x_1$. Notice that ξ is not transitive. A *tolerance space* (X, ξ) is a set X with a tolerance on it.

Example 1. X = the set of points on this page, and ξ is the tolerance given by $x_1 \sim x_2$ if x_1 and x_2 are less than a millimetre apart.

Example 2. X = the set of atoms in the paper of which this page is made, and again let ξ be the pairs less than a millimetre apart.

Example 3. If (X, ξ) is a tolerance space and Y a subset of X, then the restriction of ξ to pairs of points in Y makes Y into a tolerance space, which we denote by (Y, ξ).

Example 4. Suppose (X, ξ) is a tolerance space. Let 2^X denote the set of subsets of X. Construct a tolerance 2^ξ on 2^X by defining $A \sim B$ in 2^X if every point of A is within ξ-tolerance of some point of B and vice versa. This construction enables us to pass from the tolerance on the set of points in the visual field to a tolerance on the set of black and white pictures in the visual field.

Example 5. Suppose (X, ξ), (Y, η) are tolerance spaces; then $\xi \times \eta$ gives a tolerance on the Cartesian product $X \times Y$.

Example 6. Suppose (X, ξ), (Y, η) are tolerance spaces. Let Y^X denote the set of maps from X to Y. Construct a tolerance η^ξ on Y^X by defining $f \sim g$ if their graphs in $X \times Y$ are within tolerance under $\xi \times \eta\xi$ This construction enables us to pass from the tolerance space (X, ξ) of the visual fields and the tolerance space (Y, η) of colours to the tolerance space $Y^X, \eta^\xi)$ of coloured pictures.

Example 7. Let X be the set of two-dimensional projections of a three-dimensional world. Let ξ be the tolerance given by small parallax movements of the head. Then (X, ξ) is a rather subtle form of three-dimensional geometry. This is the geometry that is presented to the retina and from which we obtain our true three-dimensional visual perception.

Example 8. Let $f: X \to Y$ be a map and (Y, η) a tolerance space. Construct a tolerance ξ on X by defining $x_1 \sim x_2$ if $fx_1 \sim fx_2$ under η. We call ξ the *pull-back* of η and write $\xi = f^{-1}\eta$. The tolerance of visual acuity in the visual field is the pull-back of the tolerance on the retina given by the receptor fields of the retinal ganglia.

Definitions. Let $f: X \to Y$ be a map between two tolerance spaces (X, ξ) and (Y, η). We say f is a *tolerance map* if $x_1 \sim x_2$ in ξ implies $fx_1 \sim fx_2$ in η. If also the converse holds, $fx_1 \sim fx_2$ implies $x_1 \sim x_2$, we call f a *tolerance embedding*. If, further, every $y \in Y$ lies within tolerance of some fx, $x \in X$, we call f a *tolerance homeomorphism*.

Example 9. Let (X, ξ) be as in Example 2, the atoms in the page, and let (Y, η) be as in Example 1, the points on the page. Let $f: X \rightarrow Y$ map each atom to its position. Then f is a tolerance homeomorphism. This is a nice illustration of how the real page = the theoretical page to within tolerance.

Example 10. The pull-back tolerance of Example 8 makes f into a tolerance embedding. Hence to within tolerance the visual experience is embedded in the retinal experience.

Example 11. If $f: (X, \xi) \rightarrow (Y, \eta)$ is a tolerance embedding, then $(X, \xi) \rightarrow (fX, \eta)$ is a tolerance homeomorphism. Under a tolerance homeomorphism all global properties of the tolerance space are preserved.

Example 12. The composition of two tolerance maps or embedding is another. However, the composition of two tolerance homeomorphisms is not in general another; to make it into a tolerance homeomorphism we have to double the tolerance. From the technical point of view this is a nice illustration of the type of non-transitivity that occurs in the theory of tolerance spaces. From the point of view of applications we are not particularly interested in local phenomena, and so to within the double tolerance ξ^2 there is no difference between ξ and ξ^2.

Example 13. All the functions in the earlier part of the paper we claimed were tolerance embeddings. The arguments in favour of this claim are set down in [4] and [10].

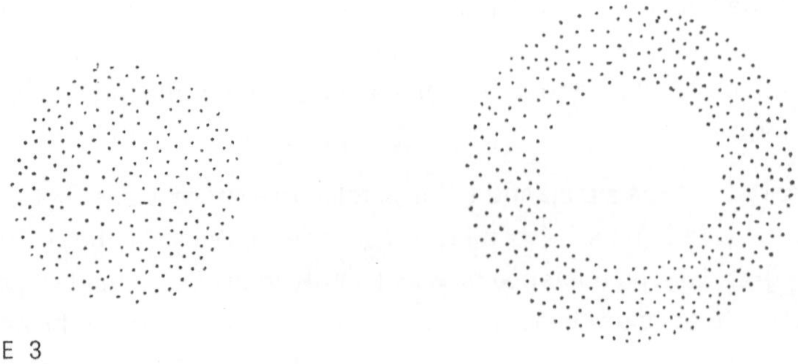

FIGURE 3

The reader will see that with a tolerance of centimetre, then the following tolerance space consists of two connected pieces, one of which has a hole in it. These are typical global properties of a tolerance space, and the best way to capture them all at once is by homology theory. Given a tolerance space (X, ξ) we can construct a simplicial complex, consisting of all simplexes, where a simplex is a finite oriented subset of X all of whose points are within tolerance of each other. The homology group $H(X, \xi)$ is defined to be the homology group of this complex. We would like the theorem:

Theorem. A tolerance homeomorphism induces isomorphisms of homology groups.

However, this is not quite true; we have to add slight technical refinements (see [11]). However, from the version of the theorem that is true we can deduce that such properties as connectedness, the number of holes, and dimension, are preserved under tolerance homeomorphism. As a consequence the rather subtle three-dimensional geometry of Example 7 is preserved through into the memory fields of the cortex, for example. Another important place where homology theory will need to be used is to discuss the coupling and resonance of the differential equations in the hippocampus and cortex.

References

1. J. C. Eccles *Physiology of Synapses* (Springer 1964).
2. W. R. Adey, C. W. Dunlop and C. E. Hendrix *Arch. Neurol. 3* (1960) 74–90.
3. W. Penfields and L. Roberts *Speech and Brain Mechanisms* (Princeton 1959).
4. E. C. Zeeman in (K. Fort, ed.) *Topology of 3 Manifolds* (Prentice Hall 1962) pp. 240–56.
5. B. G. Cragg, to appear.
6. D. H. Hubel and T. N. Wiesel, *J. Physiol. 160* (1962) 106–54.
7. P. J. van Heerden *Applied Optics 2* (1963) 387–400.
8. J. D. Green *Physiological Review 44* (1964) 561–608.
9. B. Gartside and O. C. J. Lippold *J. Physiol. 189* (1967) 475–87.
10. E. C. Zeeman *Mathematics and Computer Science in Biology and Medicine* (M.R.C. 1965) pp. 240–56.
11. E. C. Zeeman *Homology of Tolerance Spaces* (to appear).

Une Théorie Dynamique de la Morphogenèse

René Thom
Institut des Hautes Etudes, Bures s. Yvette

Le mot: morphogenèse. Selon certains puristes, le terme français Morphogenèse ne s'emploie que pour désigner l'apparition de formes organiques nouvelles au cours de l'Evolution ; en anglais le mot : «Morphogenesis» a une acception plus large, puisqu'il désigne, notamment la formation de l'organisme adulte à partir de l'embryon. Toutefois, certains auteurs anglo-saxons opposent «Morphogenesis» à «Pattern Formation» : «Morphogenesis» ne se dirait que des processus (tels la gastrulation ou la neurulation chez l'embryon d'Amphibien) accompagnés de mouvements spatiaux de caractère global ; «Pattern Formation» serait réservé aux processus formateurs à caractère statique, comme la formation des os, ou la croissance des poils ou des plumes sur la peau. Cette distinction peut sembler assez arbitraire, et d'un intérêt discutable. Ici nous emploierons le terme «Morphogenèse», conformément à l'étymologie, au sens le plus général, pour désigner tout processus créateur (ou destructeur) de formes ; on ne se préoccupera ni de la nature (matérielle ou non) du substrat des formes considérées, ni de la nature des forces qui causent ces changements.

▸ *Origine et champ d'application de la theorie.* La théorie que je propose ici provient de la conjonction de deux sources : d'une part, mes propres recherches, en Topologie et Analyse Différentielles sur le problème dit de la stabilité structurelle : étant donnée une «forme» géométriquement définie par le graphe d'une fonction $F(x)$ par exemple, on se propose de savoir si cette fonction est «structurellement stable», c'est à dire, si, en perturbant la fonction F suffisamment peu, la fonction perturbée $G = F + \delta F$ a encore la même forme (topologique) que la fonction F initiale. D'autre part, la lecture des traités d'Embryologie, et notamment des livres de C. H. Waddington, dont les idées de «chréode» et de «paysage épigénétique» m'ont paru s'adapter très précisément au schéma abstrait que j'avais rencontré dans ma théorie de la stabilité structurelle des fonctions et applications différentiables. C'est dire que la théorie présente un grand caractère d'abstraction et de généralité, et son champ d'application dépasse largement l'Embryologie, ou même la Biologie. De fait, j'en connais des applications en Optique Géométrique, en Hydrodynamique et Dynamique des Gaz (Singularités stables des fronts d'onde et des ondes de choc) ; d'une manière sans doute plus spéculative, mais néanmoins utile, la notion de «champ morphogénétique» s'identifie sur le plan physiologique à la notion de champ fonctionnel des physiologistes ; dans le cas particulier des activités nerveuses chez l'Homme, un *mot* peut être considéré comme un tel champ dans l'espace des activités neuroniques, et l'étude des associations «stables» de mots débouche sur une théorie géométrique du langage, de la «signification».

▸ *L'indépendance du substrat.* L'idée essentielle de notre théorie, à savoir qu'une certaine compréhension des processus morphogénétiques est possible sans avoir recours aux propriétés spéciales au substrat des formes, ou à la nature des forces agissantes, pourra sembler difficile à admettre, surtout de la part d'expérimentateurs habitués à tailler dans le vif, et continuellement en lutte avec une réalité qui leur résiste. L'idée cependant n'est pas nouvelle, et, on la trouve, formulée presque explicitement, dans le traité classique de d'Arcy Thompson «On Growth and Form» ; mais les idées de ce grand visionnaire étaient trop en avance sur leur temps pour s'imposer ; exprimées souvent d'une manière trop naïvement géométrique, il leur manquait d'ailleurs une justification mathématique et dynamique que seules les recherches récentes pourront leur donner. Pour illustrer cette relative indépendance de la morphogenèse vis-à-vis de notre connaissance du substrat, j'invoquerai les deux exemples suivants :

1. Considérons en premier lieu un œuf de grenouille, fécondé, que nous laissons se développer dans des conditions optimales ; en consultant les traités d'Embryologie, nous serons en mesure de prédire, (sauf accident, bien entendu), avec une très bonne précision, toutes les modifications subies par cet organisme, et toute la morphologie de son développement.
2. Considérons d'autre part une falaise mise à nu à une date déterminée t_0 par un glissement de terrain ; supposons connus la nature géologique de cette falaise, et tout le micro-climat local ultérieur (vents, pluis, températures, etc.). Peut-on prévoir la forme que prendra ultérieurement la falaise sous l'action des agents érosifs ?

Dans le cas 2, notre connaissance du substrat et des forces agissantes est excellente ; néanmoins, bien malin serait le géomorphologue qui pourrait prédire avec quelque précision la forme sculptée ultérieurement dans la falaise par l'érosion ; dans le premier cas, nos possibilités de prévision sont très bonnes ; et cependant, notre connaissance du substrat et des mécanismes morphogénétiques, (si l'on excepte quelques généralités biochimiques sur la synthèse des protéines) est des plus lacunaires. En morphogenèse, il n'y a donc qu'un rapport très distant entre nos possibilités de prévision d'une part, et notre connaissance du substrat d'autre part.

On m'objectera qu'ici j'ai comparé l'incomparable, en mettant sur le même plan un processus biologique d'une part, et un processus de la nature inanimée d'autre part. Précisément, cette comparaison fera sentir un point important, dont peu de gens sont conscients : à savoir que la morphogenèse en nature inanimée est moins bien connue et tout aussi peu comprise que la morphogenèse des êtres vivants ; cette dernière a attiré l'attention des biologistes depuis plusieurs siècles. Au contraire, la morphologie en Physico-Chimie a fait l'objet d'un mépris quasi permanent de la part des savants ; un problème important et typique est celui de la répartition géométrique d'une substance entre deux phases ; or, sur cette question, on ne sait que très peu de choses, et par exemple il n'existe, à ma connaissance, aucune théorie pour expliquer la croîssance dendritique des cristaux. Le dédain des physico-chimistes pour ce genre de questions s'explique aisément : c'est qu'il s'agit là de phénomènes extrêmement instables, difficilement reproductibles, et rebelles à toute mathématisation : en effet, le propre de toute forme, de toute morphologie est de s'exprimer par une discontinuité des propriétés du milieu ; or, rien ne met plus mal à l'aise un mathématicien qu'une discontinuité, car tout modèle quantitatif utilisable repose sur l'emploi de fonctions analytiques, donc continues. Si, jetant une pierre dans une mare, vous désirez savoir ce qui se passe, il vaut infiniment mieux faire l'expérience et la filmer, que d'essayer d'en faire la théorie ; les meilleurs spécialistes de l'équation de Navier-Stokes seraient certainement incapables de vous en dire plus. Il importe de combler ce retard ; mais il ne serait pas étonnant que la morphogenèse biologique, mieux connue, plus lente et plus strictement contrôlée, ne nous aide à comprendre certains phénomènes rapides et fugaces de la morphogenèse inanimée.

▸*Le déterminisme*. On sait qu'il y a, en principe, deux types de modèle mécanique : le modèle classique, déterministe, et le modèle quantique, foncièrement indéterminé, et dont le déterminisme ne s'exprime que statistiquement. On considère d'ordinaire que les phénomènes à l'échelle subatomique relèvent des modèles quantiques (et sont de ce fait indéterminés), alors que les phénomènes macroscopiques relèveraient des modèles classiques, et seraient, de ce fait, soumis à un déterminisme rigoureux. Cette vision livresque des choses est, croyons-nous, fondamentalement erronée. Je ne discuterai pas ici de l'indéterminisme quantique ; mais, à l'échelle macroscopique, de nombreux phénomènes présentent une certaine forme d'instabilité, due au fait que des symétries initiales disparaissent : ainsi un disque homogène abandonné en chute libre dans l'air à partir d'une position horizontale va décrire en tombant une spirale ; une baignoire cylindrique, remplie d'eau, lorsqu'on débouche le robinet central, se vide en présentant un mouvement de rotation du liquide dont le sens est «a priori» inconnu et imprévisible : dans tous les cas de ce genre, d'infimes variations des conditions initiales peuvent conduire à de très grandes variations de l'évolution ultérieure. Dans tous ces cas, il est possible de postuler que le phénomène est déterminé ; mais c'est là une position proprement métaphysique, inaccessible à toute vérification expérimentale. Si l'on veut se contenter de propriétés expérimentalement contrôlables, on sera amené à remplacer l'hypothèse invérifiable du déterminisme par la propriété empiriquement vérifiable de «stabilité structurelle» : un processus (P) est structurellement stable, si une petite variation des conditions initiales conduit à un processus (P') isomorphe à (P) (en ce sens qu'une petite transformation sur l'espace-temps – un ϵ-homéomorphisme, en géométrie – ramène le processus (P') en le processus (P)). Ceci nous conduit tout naturellement à la notion de «chréode» de Waddington, de «champ morphogénétique» en général ; un champ morphogénétique sur un ouvert U de l'espace-temps réside dans la donnée d'un «modèle universel» dont le processus donné est copié. Un tel processus est de ce fait structurellement stable ; il n'y a, donc, aucun mystère dans la notion de champ morphogénétique : cette notion exprime seulement le fait qu'un processus se passe conformément à un modèle donné «a priori», et ceci de manière structurellement stable. Dans tout processus naturel, on s'efforce d'abord d'isoler les parties du domaine où le processus est structurellement stable, les "chréodes" du processus, îlots de déterminisme séparés par des zones où le processus est indéterminé ou structurellement instable. Par l'introduction de modèles dynamiques, on s'efforce ensuite d'analyser chaque chréode en «chréodes élémentaires» associées à ce que j'appelle les «catastrophes élémentaires», puis à rapporter l'organisation de ces chréodes élémentaires en une figure globale stable par l'action d'une singularité implicite de la dynamique, le «centre organisateur». Quant à l'organisation des chréodes distinctes entre elles, le problème est plus complexe, puisqu'il est en principe indéterminé ; parmi toutes les configurations possibles de diverses chréodes, certaines sont plus stables que d'autres ; ce seront celles qui sont les plus chargées de «signification» ; ce problème difficile est qualitativement assez semblable à celui de déchiffrer un message en une langue inconnue. On va maintenant décrire notre modèle dynamique.

▸ *Description du modèle.* On va partir de l'interprétation biochimique* de la différenciation cellulaire. Considérons une enceinte dans laquelle on a mis k

* Cette idée d'interpréter la différenciation cellulaire en terme de «régime stable du metabolisme», d'attracteur de cinétique biochimique est attribuée souvent à Delbrück et Szilard. En fait, on la trouve énoncée – sous sa forme locale, qui est la seule correcte – dans Waddington, *Introduction to Modern Genetics*, 1940.

substances chimiques $s_1, s_2, \ldots, s_k$, de concentrations respectives $c_1, c_2, \ldots, c_k$. Par suite des réactions chimiques qui ont lieu entre ces substances, les concentrations c_i varient, selon une loi différentielle que nous écrirons (1) $dc_i/dt = X_i(c_1, \ldots, c_k)$. Nous ne chercherons pas à expliciter les seconds membres X_i à l'aide des lois de la cinétique chimique (action de masse . . etc.) ; le seul fait qui nous intéressera est le suivant : les équations (1) définissent dans l'espace euclidien à k dimensions R^k de coordonnées $(c_1, \ldots, c_k)$ un champ de vecteur X de composantes X_i. L'évolution du mélange sera décrite par le déplacement du point représentatif $c_i(t)$ le long d'une trajectoire du système différentiel défini par le système (1). En règle générale, le système va évoluer vers un état limite unique $c_i°$; néanmoins des cas sont possibles où plusieurs points limites peuvent exister, on peut même obtenir, parfois, une trajectoire fermée comme état limite, voire même une figure plus compliquée, comme une surface, ou (à un plus grand nombre de dimensions) une variété ; l'ensemble connexe de ces points limites sera appelé un *attracteur* du système (1). Etant donné un tel attracteur (A), l'ensemble des trajectoires du champ qui tendent vers A forme un domaine de l'espace R^k qu'on appelle le «bassin» de l'attracteur (A). Lorsque le système (1) admet plusieurs attracteurs disjoints, ces attracteurs sont en compétition ; dans quelques cas simples, les bassins relatifs aux différents attracteurs sont séparés par des «hypersurfaces» du type «ligne de crête», régulièrement plongées dans l'espace ; mais, dans des cas plus compliqués, mais néanmoins «structurellement stables» (résistant à de petites perturbations), les bassins relatifs à deux attracteurs (A_1), (A_2) peuvent s'interpénétrer de manière très compliquée topologiquement. En ce cas, le choix de l'évolution finale, vers (A_1) ou (A_2) est pratiquement indéterminé, et on ne peut guère qu'évaluer statistiquement les probabilités respectives de chaque issue en mesurant la densité locale de chaque bassin. Il est loisible, en ce cas, de parler d'une situation de «lutte», de «conflit» entre deux attracteurs.

Cela étant, nous allons maintenant «localiser» la construction précédente ; si (x) désigne un système de coordonnées dans le domaine U siège de notre processus, les concentrations c_i seront fonction des coordonnées (x,t) et vérifieront – en principe – un système d'équation aux Dérivées partielles de la forme :

$$\frac{\partial c_i}{\partial t}(x,t)_j = X_i(c_j, x, t) + k\Delta c_i \qquad \Delta \text{ laplacien}$$

où le terme $k\Delta c_i$ exprime l'effet égalisateur de la diffusion. En fait, dans tout ce qui suit, on supposera ce terme petit par rapport aux X_i, et on le traitera comme une perturbation des X_i ; si le champ de vecteurs X_i est structurellement stable (au moins au voisinage de l'attracteur régissant le régime local) – ceci est sans effet qualitatif sur l'évolution. A tout point x de U se trouve ainsi associé un champ de vecteurs sur R^k, autrement dit un «système dynamique» ; la donnée, en tout point x de U, d'un tel champ de vecteurs constitue ce qu'on appellera un «champ de dynamiques locales» sur U. On admettra qu'en tout point x la dynamique locale a déjà atteint un «état limite», un attracteur de la dynamique locale. Cet attracteur étant structurellement stable, règne sur tout un ouvert de points voisins de x ; l'ouvert U se trouve ainsi partagé en domaines U_j associés à des attracteurs A_j ; ces domaines sont séparés par des surfaces ondes de choc, qui forment ce qu'on appelle l'ensemble des *points catastrophiques* du processus. La donnée de cet ensemble constitue la *morphologie* du processus.

Il est pratiquement impossible – sans hypothèse plus précise – de spécifier la position des ondes de choc séparant les domaines des divers attracteurs ; même dans les cas théoriquement les plus simples – Hydrodynamique notamment – ce problème n'a que des solutions très partielles. Néanmoins, si l'on s'intéresse, non à l'évolution quantitative, mais seulement à la structure qualitative (topologique) des surfaces séparatrices, le problème devient plus accessible. Moyennant une hypothèse de caractère général, la «convention de Maxwell» (qui exprime en quelque sorte l'égalité des «potentiels locaux» relatifs à chaque attracteur de part et d'autre de la séparatrice) il est possible de montrer que ces surfaces séparatrices ne présentent qu'un petit nombre de singularités stables, toujours les mêmes (ceci, à tout le moins, dans le cas où la dynamique locale est une dynamique de gradient X = grad V). En ce cas, j'ai dressé la liste complète de ces singularités, qui sont autant de «catastrophes élémentaires» sur l'espace-temps R^4. En effet, ces singularités apparaissent lorsque le dynamique locale X = grad V est elle-même dans une situation «critique», par exemple lorsqu'un attracteur A est détruit, ou se divise en plusieurs attracteurs (phénomène que Henri Poincaré a appelé la «bifurcation»). On peut faire le tableau de toutes les singularités du potentiel V qui se présentent de manière structurellement stable sur R^4, et donner le modèle algébrique correspondant des surfaces de catastrophe. A titre indicatif, en voici la liste :

(i) *Le pli.* Destruction d'un attracteur, et capture par un attracteur de potentiel moindre.

(ii) *La fronce.* Bifurcation d'un attracteur en deux attracteurs disjoints. Ceci engendre en Hydrodynamique ce qu'on appelle la catastrophe de Riemann-Hugoniot (formation d'une onde de choc à bord libre).

(iii) *La queue d'aronde ou crunode.* Une surface «front d'onde» se creuse en un

sillon dont le fond est le bord d'une onde de choc. Le blastopore dans la gastrulation des Amphibiens en fournit un exemple probable en Embryologie.
(iv) *Le «papillon»*. Cette singularité du sixième ordre en V se traduit par l'exfoliation, le «cloquage» d'une onde de choc à bord libre.
(v) *L'ombilic hyperbolique*. Il s'agit de la singularité présentée par le crêt d'une vague sur le point de déferler.
(vi) *L'ombilic elliptique ou le «poil»*. Cette singularité se présente comme l'extrémité d'un «piquant», sorte de pyramide effilée à base triangulaire.
(vii) *L'ombilic parabolique*. Transition entre ombilic elliptique et hyperbolique ; il se manifeste sous la forme en champignon fréquemment présentée par un jet qui brise.
Ces trois dernières singularités sont associée à des singularités du potentiel V d'un type plus compliqué (point de «corang» deux) ; elles dirigent, en Hydrodynamique, la morphologie du déferlement ; en Biologie, très vraisemblablement, elles dirigent l'organogenèse des processus de capture (phagocytose chez les Unicellulaires) et de la sexualité (formation et émission des gamètes).

Dans tous ces cas, le schéma géométrique est le suivant : en un point (x,t) de U, la dynamique locale présente une singularité d'un type algébrique donné (s) ; on applique alors la propriété suivante, d'Analyse Différentielle : toute déformation de la dynamique critique (s) est en correspondance avec un point d'un espace euclidien W, associé à la singularité (s) ; cet espace, qui paramétrise toutes les déformations possibles (à un homéomorphisme près) de la dynamique singulière (s), constitue ce que j'appelle «le déploiement universel» de la singularité (s) ; l'évolution ultérieure du processus à partir d'une situation initiale où la dynamique locale est dans l'état critique (s) en (x_0) est alors qualitativement définie par une application F_t de U dans W (l'onde de croissance) : les catastrophes ultérieures sont alors définies par l'intersection de l'image $F_t(U)$ avec un «ensemble catastrophique universel» associé à la singularité (s) dans (W). On trouve ainsi une justification mathématique à l'idée de «paysage épigénétique» (epigenetic landscape) proposée par Waddington il y a une vingtaine d'années. En poussant ce modèle à l'extrême, on pourrait dire que l'organisme adulte n'est qu'une portion du «déploiement universel» de la dynamique germinale qui règne sur l'œuf. Les «catastrophes élémentaires» dont on a donné la liste correspondent à des singularités de la dynamique de codimension 4 : l'espace W associé est de dimension inférieure à quatre ; ce sont les seules singularités (associées à des dynamiques de gradient) qui peuvent se présenter de manière stable dans notre espace-temps ; aussi les trouve-t-on aussi bien en nature inanimée que chez les êtres vivants.

Mais il est clair qu'elles ne sauraient suffire à rendre compte de tout le développement d'un être vivant ; un premier problème réside dans la configuration stable présentée par des «chréodes» à priori indépendantes entre elles : on peut s'expliquer – parfois – cette association par l'existence d'une singularité «centre organisateur» de codimension supérieure à quatre, qui n'est pas recontré par l'onde de croissance $F_t : U \rightarrow W$; mais alors une telle évolution exige déjà la mise en route de mécanismes homéostatiques qui maintiennent l'«onde» F_t dans une région bien déterminée de (W) ; de fait, c'est bien ce que montre le développement de l'embryon : alors qu'en épigenèse primitive on ne trouve que des singularités du type «catastrophe élémentaire», les organogenèses plus raffinées, qui, comme celles de l'œil ou des os, exigent un certain contrôle métrique, n'apparaissent que beaucoup plus tardivement.

De plus, les singularités associées à des dynamiques de gradient sont d'un type très simple, et il est certain que, même en nature inanimée, on peut avoir des dynamiques locales qui présentent de la «récurrence» (trajectoires fermées ou «presque fermées»). Malheureusement, l'étude mathématique des «bifurcations» présentées par ces attracteurs pluridimensionnels, et la nature topologique des catastrophes qui en résultent, est à peine abordée (elle est d'ailleurs d'une grande difficulté). Mais une chose est certaine : alors que les catastrophes associées aux dynamiques de gradient sont définies par des ensembles de type polyèdral, les catastrophes associées à la diminution de dimension d'un attracteur (due à un phénomène de résonance, par exemple) conduisent en général à des ensembles catastrophiques d'une très grande complexité topologique, du type des formes arborescentes ou dendritiques ; c'est là l'origine dynamique commune aux formes dendritiques observées en nature inanimée (solidification : croissance des cristaux) et en nature vivante (arbres ; circulation sanguine ; les «schémas classificatoires» observés dans l'organisation des souvenirs répondent aussi à la même nécessité). De manière générale, l'apparition d'une nouvelle «phase» dans un milieu initialement homogène conduit à ce genre d'apparence, que nous appelons «catastrophe généralisée» ; tout processus dans lequel il y a rupture d'une symétrie initiale est de ce fait, structurellement instable, et conduit à une catastrophe généralisée ; de tels processus ne sont pas formalisables ; mais il faut observer, que même si le processus lui-même est structurellement instable, son issue finale, elle, peut être bien déterminée. On s'explique ainsi que la vie use couramment de catastrophes généralisées en Embryologie (qu'on compare la gastrulation des Amphibiens, suite de catastrophes ordinaires, à la gastrulation chez les Oiseaux ou les Mammifères, catastrophe généralisée). La mort d'un être vivant se manifeste par le fait que la dynamique de son métabolisme local passe d'une configuration récurrente à une configuration de gradient : c'est, typiquement, une catastrophe généralisée.

▸ *Le controle expérimental.* Nous ne pousserons pas plus avant la description de notre modèle, ce qui exigerait un développement technique considérable sans apporter de grandes clarifications sur l'essentiel. J'aborderai une question, qui, évidemment, est sur toutes les lèvres : Ces modèles sont-ils susceptibles d'un contrôle expérimental ? Il me faut, au moins pour le moment, répondre à cette question par la négative.

En effet, soit (P) le processus naturel étudié ; deux cas sont possibles : ou (P) est structurellement stable, et contenu tout entier dans une «chréode» ; en ce cas, (P) admet un modèle qualitatif donné une fois pour toutes, et on voit difficilement ce que pourrait apporter l'expérience, sinon confirmation de la stabilité structurelle de la chréode ; sans doute, on peut s'efforcer d'aborder l'étude interne de la chréode : en la décomposant en «catastrophes élémentaires», puis en rapportant la configuration des catastrophes élémentaires à l'action d'un «centre organisateur» (éventuellement extérieur au support de la chréode), on peut s'efforcer d'analyser les processus dynamiques qui en assurent la stabilité. Mais cette analyse est souvent arbitraire ; elle conduit souvent à plusieurs modèles entre lesquels on ne peut choisir que pour des raisons d'économie ou d'élégance mathématique. Par ailleurs, la théorie des catastrophes n'est pas assez avancé pour permettre l'édification d'un modèle quantitatif : le seul cas connu est celui de la catastrophe élémentaire définie par l'évolution d'une onde de choc en Dynamique des fluides : ce seul exemple montre les difficultés du problème.

Ou le processus (P) est structurellement instable, et contient plusieurs chréodes (par exemple deux chréodes (C_1), (C_2) séparées par des zones d'instabilité ou d'indétermination) ; en ce cas, on pourrait en principe rapporter l'indétermination à l'effet d'une catastrophe généralisée, non formalisable en elle-même –. Le seul espoir d'en tirer un modèle est de faire une statistique : on devra considérer, non plus un seul processus, mais tout un ensemble de processus sur lequel on fera une statistique des apparences morphologiques. (C'est la méthode suivie, d'ailleurs en Biologie Quantitative.) Mais ici encore, la théorie des catastrophes généralisées n'est pas assez avancée pour permettre la construction d'un modèle. Je crois, personnellement, que la mécanique quantique usuelle n'est – dans cette optique – qu'une statistique de catastrophes hamiltoniennes.

Devant ce constat d'impuissance à contrôler notre modèle par l'expérience, les esprits strictement empiricistes – à la Bacon – seront tentés de le rejeter comme vaine spéculation. Sur le plan de l'édification de la science actuelle, je ne peux que leur donner raison. Mais, à plus longue échéance, il y a, me semble-t-il, deux raisons qui devraient inciter tout savant à lui accorder quelque crédit :

1. La première est que tout expérimentateur travaille dans une spécialité donnée ; nous admettons comme données «a priori» ces découpages, cette taxonomie de l'expérience phénoménale en grandes disciplines : Physique, Chimie, Biologie... Or, d'où provient cette division de l'expérience, sinon d'une décomposition de notre champ perceptif en «chréodes» apparemment disjointes ? C'est bien en vain qu'on opposerait à notre modèle qualitatif les modèles quantitatifs, considérés comme seuls scientifiques et utiles. Car *tout modèle quantitatif présuppose un découpage qualitatif de la réalité*, l'isolement préliminaire d'un «système» considéré comme stable et expérimentalement reproductible. Les modèles statistiques eux-même présupposent la définition d'«ensembles» de processus stables et reproductibles. Cette décomposition que notre appareil perceptif nous lègue presque inconsciemment, tout savant l'utilise – malgré qu'il en ait – tout comme Monsieur Jourdain faisait de la prose sans le savoir. N'y aurait-il pas intérêt, dans ces conditions, à remettre en question cette décomposition, et à l'intégrer dans le cadre d'une théorie générale et abstraite, plutôt que de l'accepter aveuglément comme une donnée irréductible de la réalité ?
2. La deuxième raison est que le but ultime de la science n'est pas d'amasser indistinctement les données empiriques, mais d'organiser ces données en structures plus ou moins formalisées qui les subsument et les expliquent. Dans ce but, il faut avoir des idées «a priori» sur la manière dont se passent les choses, il faut avoir des modèles. Jusqu'à présent, la construction des modèles en Science a été avant tout une question de chance, de «lucky guess». Mais le moment viendra où la construction des modèles elle-même deviendra, sinon une science, du moins un art ; ma tentative, qui consiste à essayer de décrire les modèles dynamiques compatibles avec une morphologie empiriquement donnée, est un premier pas dans l'édification de cette «Théorie générale des Modèles» qu'il faudra bien construire un jour.

J'ajouterai, à l'usage des esprits soucieux de philosophie, que notre modèle offre d'intéressantes perspectives sur le psychisme, et sur le mécanisme lui-même de la connaissance. En effet, de notre point de vue, notre vie psychique n'est rien d'autre qu'une suite de catastrophes entre attracteurs de la dynamique constituée des activités stationnaires de nos neurones. La dynamique intrinsèque de notre pensée n'est donc pas fondamentalement différente de la dynamique agissant sur le monde extérieur. On s'expliquera ainsi que des structures simulatrices des forces extérieures puissent par couplage se constituer à l'intérieur même de notre esprit, ce qui est précisément le fait de la connaissance.

Dans le même ordre d'idées, on va montrer comment notre modèle permet d'envisager un vieux problème, celui de la finalité biologique

▶ *La finalité en biologie.* Il est à peu près admis actuellement qu'il n'y a pas d'«état vivant de la matière» : la vie ne peut être indéfiniment divisée, la cellule constituant – comme il est bien connu – l'unité irréductible de matière vivante : on est donc amené à faire de la vie une structure globale, exprimée par la présence simultanée de sous-systèmes élémentaires en une configuration (spatiale et biochimique) cohérente et stable. Cette configuration devra posséder la propriété de stabilité structurelle, et jouer le rôle de «centre organisateur» par rapport aux différentes sous-chréodes qui régissent l'évolution des systèmes élémentaires. Ainsi, il est légitime de dire – conformément au point de vue vitaliste (à la Driesch) – que tout microphénomème intérieur à l'être vivant a lieu conformément au «plan», ou au «programme» global ; mais il est non moins correct d'affirmer que l'évolution de tous ces sous-systèmes s'effectue uniquement sous l'action d'un déterminisme local – en principe réductible aux forces de la Physico-Chimie. Ainsi le point de vue vitaliste et le point de vue «réductionniste» ne sont nullement incompatibles (et des deux points de vue, contrairement aux apparences, c'est le point de vue réductionniste qui est «métaphysique», parce qu'il postule une réduction à la Physico-Chimie qui n'est pas établie expérimentalement).

Je crois à la légitimité des affirmations finalistes en Biologie ; il est vrai de dire – comme Voltaire l'affirmait en son temps – que nos yeux sont faits pour voir, et nos jambes pour marcher. Quel sens peut-on donner à ce genre d'affirmations ? Seule une analyse dynamique du développement embryonnaire permet, je crois, de préciser la signification de ces phrases. A titre éminemment spéculatif, je me hasarderai à présenter ici une telle analyse.

L'idée fondamentale de notre modèle est que toute spécialisation cellulaire étant caractérisée par un régime stable du métabolisme local, c'est à dire un attracteur A de la cinétique biochimique tangente au point considéré, la signification fonctionnelle du tissu correspondant s'exprime dans la structure géométrique ou topologique de cet attracteur A. Nous allons préciser par des exemples. Mais, auparavant il nous faut définir une notion : la figure de régulation globale d'un être vivant. Un animal, par exemple, se caractérise par sa stabilité globale : soumis à un choc, à un stimulus (s), il y répondra par un réflexe (r) qui aura pour effet – en principe – d'annuler la perturbation causée par le stimulus (s). Or les stimuli, du seul fait qu'ils proviennent de l'espace ambiant, forment un continuum géométrique pluri-dimensionnel (W). L'origine (O) de cet espace euclidien (W) désignera l'état de repos, non exicité, de l'animal ; s'il y a une infinité continue de stimuli, il n'existe, par contre, qu'un nombre fini de réflexes correcteurs (r_j) (en principe) ; ceci veut dire que si l'animal est soumis à un stimulus (s), le point représentatif de son état va en un point (s) de (W), puis revient en 0, mais il y revient en empruntant une courbe bien définie caractéristique du réflexe correcteur r(s) ; l'espace (W) des stimuli est ainsi partagé en «bassins d'attraction» associés à chacun des réflexes r_j ; c'est cette configuration qu'on appellera la «figure de régulation globale» de l'animal considéré.

Considérons maintenant un œuf de l'espèce en question ; avant fécondation, le métabolisme est faible, et est caractérisé par un attracteur de faible dimension (v_o) ; mais la fécondation amène la mise en route d'un grand nombre de cycles de réactions, le déblocage d'un grand nombre de degrés de liberté, en sorte que la dimension de l'attracteur (v_o) augmente et devient une «variété» (V) d'assez grande dimension ; c'est le phénomène de «catastrophe silencieuse», sans effet morphogénétique immédiat, mais qui s'exprime en Embryologie par le «gain de compétence». Notre hypothèse fondamentale est la suivante : sur l'ectoderme de jeune gastrula, cet attracteur (V) n'est pas lui-même fixé, mais subit de nombreuses fluctuations, entre des états (s) à grande dimension et des états (r) de dimension moindre ; par ailleurs, la topologie d'espace fonctionnel des états (s) et des états (r) en lesquels ils se dégradent est telle qu'elle réalise un modèle de la figure de régulation globale de l'espèce, notamment par l'association (s)→r(s). Cela étant, cette figure est trop complexe pour être stable ; certaines cellules se spécialisent en perdant les états (s) et ne gardent que les états (r) ; c'est d'abord l'endoderme (qui ne conserve que les états (r) relatifs aux réflexes alimentaires) ; puis le mésoderme, qui ne conserve que les états (r) relatifs aux réflexes de mouvement et de régulation biochimique. D'autres cellules, au contraire, perdent les états (r) pour ne conserver que les états (s) : ce sont les cellules nerveuses ; en effet, les cellules nerveuses, ayant perdu la capacité de réguler leur métabolisme, conservent la trace de tout ce qui leur arrive, qualité très précieuse pour le futur organe de la mémoire. (En fait, bien entendu, la régulation a lieu, mais de manière catastrophique et indifférenciée, par décharge de l'influx nerveux.) D'autres cellules (celles de l'épiderme) évoluent par vieillissement vers un attracteur situé à mi-distance entre (s) et (r), et perdent la compétence. L'attracteur (M) du mésoderme contient le groupe des déplacements euclidiens (D) ; par une dégénérescence ultérieure, certaines cellules perdent ce groupe complètement : elles deviendront des cellules osseuses (ostéoblastes) ; dans d'autre cas, la dégénérescence sera moins complète, et l'attracteur contiendra encore un sous-groupe à un paramètre du groupe (D) : elles deviendront des cellules musculaires (myoblastes), et la géométrie de cette dégénérescence, transposée dans une «chréode» à caractère métrique, décrit la formation des os et des muscles qui leur sont attachés. La formation des organes sensoriels est très analogue : le champ (s) se décompose en un produit direct de champs (S_v) (S_a) (S_t), etc., visuel, auditif, tactile...et chacun de ces champs l'emporte de manière définitive sur une zone préférentielle du tissu neural ; dans le cas du champ visuel (S_v) par exemple, le groupe (D) des déplacements opère également dans (S_v) ; une dégénérescence ultérieure conduit à la formation d'une «chréode» métriquement contrôlée, qui est le glove oculaire ; l'action du groupe (D) se manifeste par la présence de mésenchyme dans cette chréode qui aboutira à la formation de la choroide et de la sclérotique, sur laquelle vont s'attacher les trois paires de muscles qui «symbolisent» la décomposition en trois sous-groupes à un paramètre du domaine de (D) qui agit sur (S) ; cette action du groupe (D) va se retrouver, de manière compensatrice, dans le champ des activités neuroniques stationnaires en lesquelles le champ (S_v) se reconstitue approximativement, une fois terminée l'organogenèse du cerveau. C'est là, en effet, un fait général : un champ régulateur ($s \to r(s)$) subit au cours de l'organogenèse toute une série de décompositions, de «catastrophes», dues à la formation d'organes partiels ; mais, finalement, le champ se reconstitue en tant que forme globale d'activités nerveuses. Ainsi, par exemple, un réflexe alimentaire type (F) met en action : (1) un stimulus : la vision d'une proie (p) (2) Un champ moteur (r) : saisir la proie l'amener à la bouche, la manger (3) Un champ viscéral, dirigeant les activités motrices et glandulaires du tube digestif. Je considère comme probable qu'un tel champ existe déjà sous forme de transformations géométriques préférentielles $s \longleftrightarrow r(s)$ dans le métabolisme de l'ectoderme de la jeune gastrula ; puis il se localise de manière plus importante dans l'endoderme, mais en subsistant sporadiquement dans les autres tissus. Lorsque vient le moment de la formation de la bouche et des dents, cette formation est induite par le contact de l'endoderme avec l'ectoderme et le mésenchyme issu des crêtes neurales : le champ (F) se reconstitue alors par résonance dans les tissus compétents en contact avec l'endoderme, où il engendra la bouche et les dents. La «chréode» de l'organogenèse apparait ainsi comme l'arête de départ, le «bord libre» de l'«onde de choc» physiologique qui définit la mise en activité du champ (F). Cette conception explique – dans une certaine mesure – la formation des Chimères Têtard-Triton dans l'expérience classique de Spemann, où de l'Ectoderme greffé de Triton concourt à la formation de la bouche de l'hôte avec ses propres moyens génétiques : il y a certainement un isomorphisme grossier entre les «figures de régulation» d'espèces animales, mêmes assez lointaines dans la phylogenèse, parce que les contraintes imposées par la régulation, l'homéostasie des animaux sont grosso modo les mêmes pour tous, et compensées par les mêmes fonctions ; c'est dans le détail des catastrophes de l'organogenèse que ces champs engendrent des structures variables et différentes selon l'espèce. La reconstitution – sous forme d'activités nerveuses canalisées en «chréodes»–de champs fonctionnels décomposés par les catastrophes de l'organogenèse – peut apparaître comme un processus mystérieux d'inspiration vitaliste ; je dispose cependant d'un schéma dynamique abstrait qui peut en rendre compte, au moins sous forme théorique. Il existe ainsi, entre le cerveau et la gonade, une certaine homologie fonctionnelle : le cerveau (plus généralement le système nerveux) reconstitue sous forme d'activités nerveuses stables les champs fonctionnels primitifs ; dans la gonade se reconstitue, en chaque gamète, le «centre organisateur» de la dynamique globale de l'espèce, la «figure de régulation» spécifique. (En fait, avec le blocage du métabolisme qui intervient à la fin de la gamétogenèse, la figure globale se «cristallise» en quelque sorte en structures macromoléculaires dont les «vibrations» biochimiques pourront reconstituer la figure totale après fécondation.)

Il est superflu d'insister sur le caractère hypothétique de cette présentation de la dynamique globale d'un animal. Ma seule ambition est ici de fournir un cadre conceptuel acceptable pour s'expliquer une question obscure et complexe.

CONCLUSION

La synthèse ainsi entrevue des pensées «vitaliste» et «mécaniste» en Biologie n'ira pas sans un profond remaniement de nos conceptions du monde inanimé. On use sans trop de scrupules en Biologie (et surtout en Biologie Moléculaire !) de vocables anthropomorphes tels que : information, code, message, programme... En pure Physico-Chimie, l'usage de ces vocables serait considéré comme la manifestation d'un anthropomorphisme délirant. Notre modèle attribue toute morphogenèse à un conflit, à une lutte entre deux ou plusieurs «attracteurs» ; il apparait ainsi comme un retour aux idées (vieilles de 2500 ans !) des premiers Présocratiques, Anaximandre et Héraclite. On a taxé ces penseurs de «confusionisme primitif» parce qu'ils utilisaient des vocables d'origine humaine ou sociale comme le conflit, l'injustice pour expliquer les apparences du monde physique. Bien à tort selon nous, car ils avaient eu cette intuition profondément juste : les situations dynamiques régissant l'évolution des phénomènes naturels sont fondamentalement les mêmes que celles qui régissent l'évolution de l'homme et des sociétés, ainsi l'usage de vocables anthropomorphes en Physique est foncièrement justifié. Dans la mesure où l'on fait du «conflit» un terme exprimant une situation géométrique bien définie dans un système dynamique, il n'y a aucune objection à user de ce terme pour décrire rapidement et qualitativement une situation dynamique donnée. Qu'on géométrise de même les termes d'«information», de «message», de «plan» (ce que s'efforce de faire notre modèle) et toute objection à l'usage de ces termes tombera. La Biologie actuelle fait de la sélection naturelle le principe exclusif – le deus ex machina – de toute explication biologique ; son seul tort, en l'espèce, est de traiter l'individu (ou l'espèce) comme une entité fonctionnelle irréductible : en réalité la stabilité de l'individu, ou de l'espèce, repose elle-même sur une compétition entre «champs», entre «archétypes» de caractère plus élémentaire, dont la lutte engendre la configuration géométrique structurellement stable qui assure la régulation, l'homéostasie du métabolisme, et la stabilité de la reproduction. C'est en analysant ces structures sous-jacentes plus profondément cachées, qu'on parviendra à une meilleure compréhension des mécanismes qui déterminent la morphogenèse de l'individu et l'évolution de l'espèce. La «lutte» a lieu, non seulement entre individus et espèces – mais aussi, à chaque instant, en tout point de l'organisme individuel. Rappelons ce qu'a dit Héraclite : Il faut savoir que le conflit est universel, que la justice est une lutte, et que toutes choses s'engendrent selon la lutte et la nécessité.

Correspondence between Waddington and Thom

1. Waddington to Thom

Edinburgh, 25 January 1967

. . . There is one point in your M S I want to comment on – at first sight a personal one, but actually it is not quite trivial. You refer on page 7 to 'l'interprétation biochimique (due à Delbrück et Szilard) de la différentiation cellulaire'. I know that this interpretation – in terms of 'alternative steady states' – is usually attributed to them, Delbrück in 1949, I think, and Szilard a bit later. But actually I had stated the main point as early as 1939, but I did so (a) in a few sentences in a textbook, *Introduction to Modern Genetics*, and people are not ready to admit that there can be new ideas in textbooks, and (b) I got it right, and spoke of alternatives between time-extended chreods (though I did not yet call them that), whereas Delbrück and Szilard had the simpler and basically inadequate idea in the context of development, of an alternative between steady states. I should therefore like you to add my name on this page, writing 'due à Waddington, Delbrück et Szilard)'. You refer to the subject again on page 18, and there the situation is more difficult. You write '. . . toute spécialisation cellulaire étant – selon l'idée de Delbrück et Szilard – caractérisée par un régime stable du métabolisme, c'est à dire un attracteur A de la dynamique biochimique locale. . . .' Now the unsophisticated biologist will interpret 'régime stable' to mean an unchanging regime, a steady state. But, of course, your model does not require anything so limiting. I wonder if you could rewrite this sentence to indicate that the 'régime stable du métabolisme' may be changing in a defined way with time?

2. *Thom to Waddington* (translated from the French)

Bures, 27 January 1967

. . . I come now to the question of the reference to Delbrück and Szilard; I admit that I do not know at all where to find this reference and so I have simply to follow what is said in all the articles on this subject. Have they really believed, as you state, that a cellular differentiation is to be attributed to a definitive choice between two 'stable regimes', defined *in abstracto*, and indefinitely fixed? It is possible, particularly if their knowledge in biology was not very well developed. They were, however, physicists to whom the distinction between the local state and the global state should have been familiar. But in any case that is a purely academic question if you had made the same point before them. I therefore propose, on page 7, line 7 of my MS to suppress the parenthesis '(due to Delbrück and Szilard)' and to introduce a footnote along the following lines: 'The idea of interpreting cellular differentiation in terms of "a stable regime of the metabolism", i.e. of an attractor of the biochemical kinetics, is often attributed to Delbrück and Szilard. In fact it was stated – under its local form, which is the only correct one – in C.H. Waddington (*Introduction to Modern Genetics*)'.

On page 17 the simplest thing to do is to suppress the reference to Delbrück and Szilard and add *local* after *metabolism*.

Your point (b) brings to mind a question that seems *a priori* quite important. You are the author of the word 'chreod' and you therefore have the right to define its sense; now I have been using it, and, I believe, in a sense which is much more general than yours. I have the impression that, for you, chreod can be identified with 'developmental pathway' in the sense that a chreod will be associated to an attractor of the local biochemical kinetics. It would have no sense, in that case, to consider chreods in which several attractors might be in competition. After there had been a 'switching point' (bifurcation, in my terminology) there would be nothing left of the chreod, properly speaking. Further, since the organogenesis of the greater part of the organism requires the interaction of tissues of different types (for example an epithelium and a mesenchyme) controlled by different attractors, the formation of such an organ such as the kidney, for example, would never be described by a single chreod.

Personally, I have a tendency to employ chreod as a synonym for 'support of a morphogenetic field', without restricting the number of attractors in competition or the morphology of the domains which they control within this support. I should be glad if you would tell me if you approve of this use of the word chreod. If not, I'll have to give it up and use only the term 'morphogenetic field'.

3. Waddington to Thom

Edinburgh, 4 February 1967

Thanks for your letter of January 27th. I think your suggestions about how to deal with the Delbrück - Szilard references are probably quite all right. I confess, however, that I am not very sure of the distinction you are making between 'état locale' and 'état globale'. The distinction I want to make is between a regime (flux equilibrium) which remains unchanging throughout a period of time and a regime which is, at any moment, stable, but which changes progressively as time passes. Thus when you suggest that your sentence on page 18 should read '. . . régime stable du métabolisme locale', I feel I should like to say something like '. . . régime stable mais évoluante du métabolisme locale'. But my grasp of French is very weak, and this may not be necessary.

I am sending you Xeroxes of Delbrück's remarks which are the source of the reference to him. I am not sure where the Szilard reference is to be found. I also send you a Xerox of the pages in my *Introduction*. You will see from the bottom of page 181 that I knew I was trying to talk topology but had not the technical training to do so (see pp. 169–196).

About the meaning of 'chreod'. I think your expanded use of it is absolutely justified; indeed, I would have enlarged it in a similar way myself, except that I have always been trying to get the basic idea across to an audience of biologists who could hardly understand even the simplest applications of it. I have, however, written about the ectoderm of an amphibian gastrula being switched into the alternative chreods of epidermal, neural, or mesodermal differentiation ; and it is obvious that within each of these 'gross chreods' there are a number of sub-chreods – within the mesodermal, for instance, there will later be switching into muscular, mesenchymal, glandular-epithelial, and so on ; and that there will be interactions between the regions of tissue which have taken one chreod with those which have taken others. I think my usage has therefore implied that a chreod such as 'the mesodermal' may at a later stage in time come to include a number of different (and interacting) attractors (such as those for muscle, dermis, connective tissue, etc.). What I am not sure of is whether, at the time of a catastrophe which brings about a bifurcation into chreods A and B (neural and mesodermal, let us say), there can in any sense already be more than one attractor included with each of the chreods A and B – or do these contain only one attractor at that time, which either is inherently unstable so that it gives rise later to two or more attractors, or is unstable with respect to influences impinging on it from geographically other parts of the system, which induce the appearance of new chreods ?

Annex 1. From *Unités biologiques douées de continuité génétique* (C R N S, Paris, 1949), p. 33
M. Delbrück. – Dans sa discussion des phénomènes observés par Sonneborn et lui-même M. Beale a proposé de considérer ces phénomènes comme résultant des propriétés d'une population de plasmagènes dont la reproduction serait sélectivement favorisée ou inhibée par les conditions de milieu.

Je n'entends pas disputer cette conception, mais je voudrais attirer l'attention sur certaines propriétés générales des systèmes dits «en équilibre de flux», propriétés que l'on doit prendre en considération avant de postuler l'existence d'unités biologiques douées de continuité génétique dans l'un quelconque ou dans tous les cas où la continuité génétique d'une fonction est observée.

L'argument que je voudrais développer est le suivant : *de nombreux systèmes en équilibre de flux sont capables de plusieurs équilibres différents dans des conditions identiques. Ils peuvent passer d'un état d'équilibre à un autre sous l'influence de perturbations transitoires.*

Cette proposition générale peut être illustrée par un modèle simple.

Dans le diagramme ci-dessous, les lettres A_1, A_2, B_1, B_2 représentent des enzymes différent, contenus dans une cellule, représentée par le cercle. Les lettres

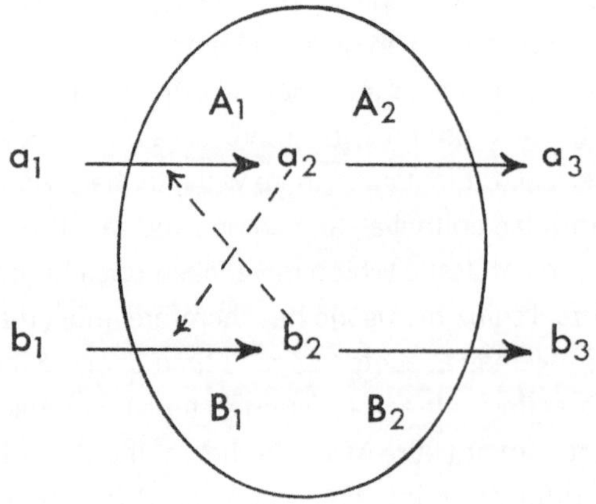

FIGURE 1

a_1 et b_1 représentent des substances du milieu. Sous l'influence des enzymes A_1 et B_1 respectivement, ces substances sont transformées en métabolites intermédiaires, a_2 et b_2. Ceux-ci à leur tour sont les substrats des enzymes A_2 et B_2 qui les transforment en produits de déchet, a_2 et b_2. Dans des conditions de milieu constantes, la cellule atteindra rapidement un état stable, caractérisé par une certaine concentration constante des produits intermédiaires a_2 et b_2. Dans ce modèle, il n'y a qu'un état stable, déterminé par les conditions de milieu et les propriétés enzymatiques de la cellule.

Ajoutons maintenant l'hypothèse qu'il existe des interactions mutuelles entre les deux séries de réactions enzymatiques. Précisons cette hypothèse en supposant que le métabolite a_2 exerce sur le taux de la réaction catalysée par l'enzyme B_1 une influence telle que pour de *fortes* concentrations de a_2, cette réaction soit inhibée.* Nous supposons d'autre part une action semblable du métabolite b_2 sur l'enzyme A_1. Ces interactions sont exprimées dans le diagramme par les flèches en pointillé.

Avec ce nouveau modèle, il est encore vrai que dans des conditions constantes, la cellule atteindra un équilibre de flux. Mais il existe maintenant trois équilibres possibles pour les mêmes conditions de milieu, deux stables et un labile. Considérons par exemple les conditions définies par l'égalité de concentration des substances a_1 et b_1. L'équilibre de flux qui se trouvera réalisé en définitive, dépendra de l'ordre dans lequel ces substances auront été ajoutées au milieu. Suivant les cas, l'équilibre sera caractérisé par :

* Une telle propriété pourrait être due à une dimérisation réversible de a_2, le dimère seul inhibant la réaction de B_1.

a. beaucoup d'a_2, peu de b_2, si a_1 a été ajouté en premier. Cet équilibre est stable, nous l'appellerons l'état *a*.

b. peu d'a_2, beaucoup de b_2, si b_1 a été ajouté en premier. Cet équilibre est également stable. Nous l'appellerons l'état *b*.

c. concentrations égales et faibles de a_2 et b_2 si les deux substances sont ajoutées ensemble en quantités égales. Il y a équilibre de flux, mais c'est un équilibre instable, dans lequel de faibles perturbations provoqueront le passage à l'état *a* ou à l'état *b*.

Le passage de l'état *a* à l'état *b* pourra être provoqué par des perturbations fortes *transitoires*. Par exemple, si l'état initial est *a*, une interruption *momentanée* de l'inhibition de B_1 par a_2 provoquera le passage de l'état *a* à l'état *b*.

Ces résultats pourront être obtenus grâce à des interventions diverses : traitement *temporaire* par un sérum anti-a_2, changement *temporaire* de température, tel que l'activité de l'enzyme A_1 soit sélectivement ralentie, transfert *temporaire* dans un milieu dépourvu de la substance a_1.

En résumé, notre modèle de cellule est capable d'exister sous deux états fonctionnellement différents d'équilibre de flux, sans que cela implique un changement quelconque dans les propriétés des gènes, plasmagènes, enzymes, ou de toutes autres unités structurales ; les passages d'un état à un autre peuvent être provoqués par des modifications *transitoires* des conditions de milieu.

Les modèles de ce type peuvent être modifiés d'une infinité de manières, afin de rendre compte d'un grand nombre d'équilibres de flux différents, doués de n'importe quel degré de stabilité. Les passages des uns aux autres pourraient être, suivant les cas, réversibles ou irréversibles, comme dans les phénomènes de différenciation, pour l'explication desquels on a également invoqué, sans preuves concrètes, l'existence de plasmagènes.

Je ne prétends pas proposer ici une théorie des phénomènes décrits par Sonneborn et Beale. Je désire simplement insister sur le fait que, dans le cas des systèmes en équilibre de flux (mais non des systèmes en équilibre) diverses explications de ce genre peuvent être envisagées, qui ne sont nullement invraisemblables, ni même improbables, d'un point de vue général. La proposition présentée plus haut n'est pas nouvelle, et de nombreux biologistes ont une notion plus ou moins claire de ce qu'elle implique. J'ai supposé que ce modèle simple aiderait à en concrétiser l'idée et à la préciser.

Annex 2. From C.H.Waddington. *An Introduction to Modern Genetics* (Allen and Unwin, London, 1939) pp. 180–4

3. *Time-effect and dose-effect curves*

In the examples mentioned in the last section we have attempted to describe the course of the developmental reactions by which a certain substance is produced. We might summarize the reactions in a given case by plotting the quantity of substance present against the time. The curve which would be obtained might be called the time-effect curve of the gene under investigation. We shall in this section try to generalize this idea so as to make the time-effect curve of a gene summarize all the information which we have about the developmental action of the gene.

In the first place, we must inquire into the relations between the time-effect curve and the dose-effect curves of the same gene. The dose-effect curves which were discussed earlier (p. 164) were obtained by plotting the dose of the gene against the final effect produced in the adult organisms. From a developmental point of view, the final effect of a gene in the adult must either be an asymptote to the time-effect curve, when development slows off gradually as maturity is attained, or in some animals it may be an end-value reached when development is suddenly brought to an end by a metamorphosis. In either case, the end-value of the time-effect curve is the same as the value plotted for that gene on the dose-effect curve. If we have a set of allelomorphs, the dose-effect curve is in fact merely a summary of the end-values of the separate time-effect curves.

The importance of this point is that it shows that certain conclusions about dose-effect curves also apply to time-effect curves. For instance, we have rather little detailed information about the dependence of the time-effect curves on the genotypic milieu, although Ford and Huxley have described the effect of some modifying factors on the time-effect curve of pigment formation in the eyes of Gammarus and a few other cases are known. But we have much more evidence from dosage compensation, cases like that of shaven, etc., which shows that the dosage curves are dependent not solely on the particular gene under investigation, but rather on the balance between that gene and the whole of the rest of the genotype. We can now see that this effect of the genotypic milieu on the dosage curve must be a result of its effect on the time-effect curves, and we can thus give a much stronger basis to the important conclusion that the time-effect curve is a function of the whole genotype.

The simplest type of time-effect curve is that in which we summarize certain developmental processes which are directly observable, such as the deposition of pigment in the eyes of Gammarus or the skins of Lymantria caterpillars. But the investigations on eye pigments in Drosophila, for instance, clearly show that the observable process is only the final reaction in a whole series of changes which lead from the gene to the pigment. We can, ideally, expand the idea of the time-effect curves to cover not only the progress of the final observable process but

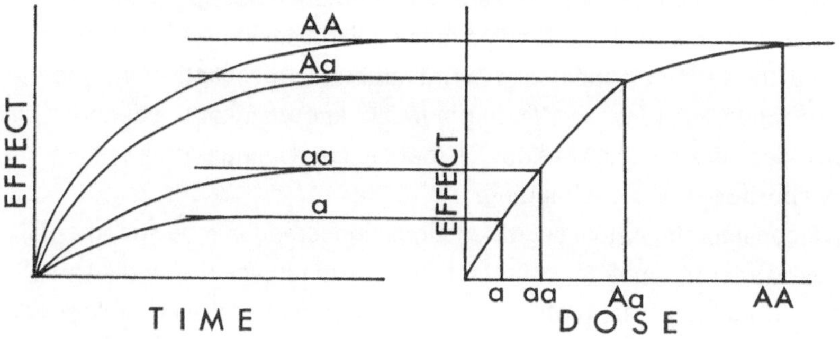

FIGURE 2
The relation between time-effect and dose-effect curves

also that of the earlier processes, about which we usually know very much less. For instance, if a pigment is formed from a precursor, we can not only plot a curve showing the speed of formation of pigment, but we can include an earlier curve which gives the rate of formation of the precursor.

If we attempted to formulate this in a strict way, we should find that for every new substance whose concentration we wished to plot we should have to introduce a new dimension of space, and this soon becomes rather alarming to non-mathematicians who are not used to creating universes to their own specifications. But even without any complication of multi-dimensional space, it is very easy to grasp the essential points which emerge when the time-effect curves are generalized in this way.

In the first place, we find cases in which the effects produced differ merely in quantity and vary continuously over a certain range; then all that we can deduce about the time-effect curves is that the rates of the processes concerned can vary continuously, so that different quantities of the end-product are produced when development ceases. Perhaps more important are the cases in which there are several fairly sharply demarcated and alternative developmental processes, which can only be represented by a system of branching lines. For instance, we have seen that in Drosophila there is a period of development when the normal vermilion substance is essential for normal eye pigmentation. If vermilion substance is absent, the pigment-forming substances will change so as to form vermilion pigment; if vermilion substance is present, they change so as to produce normal pigmentation, but come to another branching point where the presence or absence of cinnabar substance decides in which direction they will proceed. In such a case we have a mixture of reacting substances, say two or three enzymes and some substrates, and at a branching point there are two alternative possible ways in which the mixture can change, according as the vermilion substance is present or not; for instance, the vermilion substance might inhibit the most active enzyme and allow a less active one to work. We do not in fact know any of the details about the processes involved; all we know is that we are dealing with a system with alternative possible ways of changing.

If we want to consider the whole set of reactions concerned in a developmental process such as pigment formation, we therefore have to replace the single time-effect curve by a branching system of lines which symbolizes all the possible ways of development controlled by different genes. Moreover, we have to remember that each branch curve is affected not only by the gene whose branch it is but by the whole genotype. We can include this point if we symbolize the developmental reactions not by branching lines on a plane but by branching valleys on a surface. The line followed by the process, i.e. the actual time-effect curve, is now the bottom of a valley, and we can think of the sides of the valley as symbolizing all the other genes which co-operate to fix the course of the time-effect curve; some of these genes will belong to one side of the valley, tending to push the curve in one direction, while others will belong to the other side and will have an antagonistic effect. One might roughly say that all these genes correspond to the geological structure which moulds the form of the valley. Genes like vermilion which have their main effect at certain branching points are like intrusive masses which can divert the course of the developmental processes down a side valley.

This attempt to symbolize the developmental reactions may seem unduly picturesque and too abstract to be of much value. Its abstractness, however, must be blamed on the fact that we know so little about the actual processes concerned. The two important, but abstract, facts which are expressed in visual form by the valley model are, firstly, that the course of any developmental process is determined by many genes, and secondly that these genes often define alternative courses along which the reactions may go.

This same scheme may be used to describe the development of characters which are not simple substances. For instance, a similar history of successive reactions has been invoked to explain the developmental effects of the Bar gene in Drosophila which has been worked out in some detail by studying the effect of temperature on the number of facets formed in the eye. Unfortunately, we only see the end-result 'frozen' by the occurrence of metamorphosis. It is found that there is only a certain period during development (just before the facets actually appear) during which temperature changes affect the number of facets formed in Bar flies.

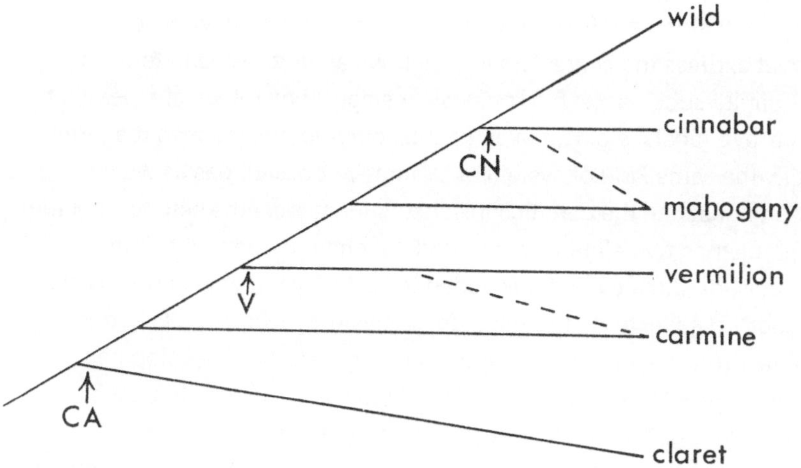

FIGURE 3

Diagram of the developmental processes of pigment formation in the eyes of Drosophila. The developmental process moves from left to right along the branching tracks. The points marked *ca*, *v*, and *cn* symbolize the alternatives dependent on whether the claret, vermilion, and cinnabar substances are produced. There must be very many more tracks which we cannot yet fit into the picture. Thus it is known that carmine has no, or little, vermilion substance, and mahogany little cinnabar substance; but we do not yet know whether the developmental processes in flies homozygous for these genes branch off from the normal track before the vermilion and cinnabar forks, or are secondary branches, as indicated by the dotted lines

The interpretation suggested is that during early development a facet-forming 'substance' is formed and that the Bar gene sets going reactions which break down this substance; then a third set of processes determine that facet formation shall actually begin, and a number of facets are formed proportional to the amount of substance still present. The temperature effects, which only occur when the Bar gene is present, presumably affect the breakdown process for which Bar is responsible. In this example the end-product which is the 'effect' in the time-effect curve is not a single substance like a pigment, but is a relatively complicated tissue, the eye facets. The facet-forming substance, however, may be a single chemical compound, since Ephrussi has recently shown that the number of facets formed in Bar eyes may be increased by the injection of suitable extracts from normal pupae. We know nothing about the mechanism of action of the substance; it might either increase some sort of inductive stimulus to facet formation or lower the threshold of reaction to such a stimulus. The substance is probably related in some way to the vermilion substance, and may be identical with it; it is known that Bar inhibits the formation of vermilion substance in the eye itself, though not in the rest of the body.

In considering development from an embryological point of view we can, as with Bar, not yet express the characters in which we are interested in terms of quantities of definite substances, but must talk instead of histological types such as neural tissue, eye facets, etc. But experimental embryology leads to the formulation of exactly the same kind of system of alternative possibilities as we have had to develop to describe the genetical results. For instance, the ectoderm of the amphibian gastrula has two alternative methods of change open to it; it may become epidermis, or, if the evocator is added to it, it may become neural tissue. The case is exactly parallel to that of the pigment system in Drosophila at one of its branch points. Both the methods of approach to the study of development formulate the main problems in the same kind of way, and we may hope that genetics and embryology can collaborate in finding the answers.

We must now consider certain genetical problems in the light of the scheme of thought which has just been developed.

4. Thom to Waddington

Bures, 20 February 1967

Many thanks for your letter of February 4th, and the enclosed texts of Delbrück and yourself; both are, I believe, of fundamental importance for the history of ideas about cellular differentiation. But I must confess that I have not been able to grasp the difference you seem to make between Delbrück's 'flux equilibria' and your own 'chreods or developmental pathways'. You seem to imply that in Delbrück's model, some alternative has been taken which, at least for some time, determines the metabolism into a fixed steady state. I do not think that Delbrück himself would have accepted this interpretation; first there is the distinction made by Delbrück: 'dans le cas des systèmes en équilibre de flux (mais non des systèmes en équilibre) . . .' (line 8 from the bottom, page 171). This seems to mean that 'flux equilibria' may move to some extent, while simple equilibria do not. But, of course, the terminology is rather obscure. Moreover, remember that this discussion took place about 'plasmagenes' or cytoplasmic heredity. Now, how could this so-called steady state survive the tremendous metabolic changes undergone by any cell in mitosis, sporulation, or gametogenesis?

When Delbrück exhibits his local model with substrates a_1, b_1, a_2, b_2, enzymes A_1, B_1, A_2, B_2, etc., quite clearly he does not claim to describe the whole metabolism of the cell. His only hope is to specify a subsystem (S) of metabolites one might reasonably consider as independent – or at least weakly coupled – to the remaining part (R) of the metabolism, involving reactions with other substances than that of the system (S). The 'stable solutions', the 'attractors' of the subsystem (S) – provided they are structurally stable – will stay qualitatively the same – or vary very little for a relatively long period of time. (Note that this observation applies to all models in biochemistry, like the regulation models à la Jacob-Monod.) But the remaining part (R) of the metabolism may itself vary in large amounts, producing very large phenotypic effects. Eventually this large variation may, through a weak coupling with (S), cause the destruction by catastrophe of the given attractor in (S), destroying the validity of the model.

Now, in your 'chreods' or developmental pathways, the only conceivable way of ensuring the stability, the 'canalization' of the chreod is to associate to the chreod a subsystem (S) of the Delbrück type. This is in fact what you express at the bottom of page 174 of your text when you refer to the sides of the valley as defined by the action of antagonistic genes (or sets of genes). Now, we have to take into account the fact that the co-ordinates of this local split factor (S) of the metabolism may not be, as in Delbrück's model, concentrations of metabolites, but very complicated functions of them (so that a simple biochemical interpretation of (S) will be difficult to explicate). Moreover, there may be, at a given time, several such split factors (S_1), (S_2), . . ., each provided with a certain set of attractors: this will give the most likely interpretation of the notion of 'homologous chreods' in embryology. Finally, I do not think that there is a fundamental difference between Delbrück's model and yours, except that yours is more general. The main merit of Delbrück, in that matter, is to have put this idea in a more specialized form, more easily acceptable to standard biochemists.

The conclusion to be drawn from this discussion is that such expressions as 'flux equilibria', 'local stable regime', 'steady state', etc., can be made precise only through the use of a geometric model – even if it is not explicit. In the correction you suggest in my text: 'régime stable, mais évoluant, du métabolisme local', I would like to say: 'un attracteur de la cinétique biochimique tangente au point considéré'. This would be the only correct statement, but who, among biologists, could understand such a language?

As I am not sure to know all your work, may I ask you to give me the references you would consider best for:

1. The word chreod.
2. The epigenetic landscape.
3. The concept of homeorhesis as opposed to homeostasis.

Many thanks in advance.

5. Waddington to Thom

Edinburgh, 23 February 1967

Many thanks for your letter. You ask: what difference do I see between what Delbrück was saying, and what I had said. Now I admit I may have misunderstood Delbrück, but what I understood him to say was as follows. He drew an enclosed space, into which there flowed certain things (labelled a_1, b_1) and out of which there came certain other things (a_3, b_3). Inside the space was his switch mechanism. When he spoke of 'systèmes en équilibre de flux (mais non des systèmes en équilibre)' in the last paragraph of his remarks, I took him to mean, by the first phrase, 'systems which are in equilibrium even though material is continually passing into and out of them, i.e. are in steady states', in contrast to the second phrase, which I took to mean 'systems in equilibrium with no flux through them'. This is, I think, the usual way in which the terms 'flux equilibrium' are interpreted in English. Now in both these phrases the meaning of 'equilibrium' – whether 'flux' or not – would normally be that the concentrations of the substance inside the box remained constant. You ask – how could they, throughout such profound changes as cell division, etc.? I should point out that the Delbrück model was proposed as an *alternative* to the postulation of genetically continuous 'plasmagenes', and was specifically designed to account for the transmission of an unchanging character through many cell generations. It was implied, I think, that the turmoil of cell division was a transient fluctuation in some system only loosely coupled with the character-preserving system he was describing.

This is, I think, the way in which most biologists have interpreted Delbrück, e.g. Jacob and Monod, 'Genetic Repression and Allosteric Inhibition', in Cyto-differentiation and Macromolecular Synthesis, 21st Growth Symposium, Acad. Press, 1963, on page 53, wrote 'Conservation implies differential functional activity of nucleotide sequences, resulting, for instance, from the establishment of steady state systems capable of clonal perpetuation, as pointed out by Delbrück (1949)'.

My point is that progressive differentiation is *not* clonal propagation.

So we come to the difference between Delbrück's system – as I interpret it – and mine. In his system the character of the cell which is in one of its alternative states remains constant in time – the concentrations of substances are in 'flux-equilibrium' in the sense that they do not change, in spite of flow through the system. This is 'homeostasis'. In my picture the alternative states of the cell are between homeorhetic 'equilibria', in any one of which the concentrations of substances do not remain constant, but change along defined time-extended trajectories.

'Who said it first?' is, of course, a pretty trivial question; but 'Who said it right?' is more important. I should claim that Delbrück in 1949 was talking about the alternatives of driving round and round the Place de la Concorde, or round and round the Etoile; and that this is only a degenerate case of what I had been talking about in 1940, which is the alternative of taking the bus from the Aérogare des Invalides to the Aéroport Orly or the Aéroport Le Bourget.

The only way to eliminate this difference between Delbrück and myself would be if you are so 'pure' a mathematician that you acknowledge no difference between a dimension devoted to a material variable, such as concentration, and one devoted to time. But this is a level of abstractness at which mathematics loses touch with the real-world problems of biologists.

If you want to give references to my discussions of these concepts, they should be:

(i) the word Chreod – *Strategy of the Genes* (Allen and Unwin: London 1957), p. 32.

(ii) The Epigenetic Landscape – *Organisers and Genes* (Cambridge University Press: 1940, and *Strategy of the Genes*.

(iii) Homeorhesis – *Strategy of the Genes*, was the first place where I proposed the word – the concept, as I have tried to show you, was implicit in the *Introduction to Modern Genetics* of 1939 (Allen and Unwin: London) and the *Organisers and Genes* of 1940.

Constants of Nature: Biological Theory and Natural Law

Paul Lieber
University of California

Introduction. The idea of constancy appears to be primitive in both scientific and mathematical thought, in so far as it is psychologically evoked as the mind attempts to organize its experience, and to formulate precise and thus definitive propositions by which the objects of its mathematical investigations are axiomatically defined. This concept has been tacitly used in scientific theory in various contexts, all of which appear to express a central philosophical principle, and have thereby consistently produced powerful and far reaching consequences.

In this paper I examine some of these cases and attempt to demonstrate thereby that each case silently invokes, and in a particular way, a central idea which unifies them and which stems from the primitive notion of constancy.

Particular attention is given to the constants of nature which have emerged in physics, as the most fundamental examples and manifestations of constancy in nature and as the ultimate foundation and expression of its laws. In this regard, attention is focused on the special theory of relativity, as it is the only physical theory which explicitly formulates, and thus explicates in terms of observables, that a particular universal constant C is a universal constant in the three-infinite set of Galilean frames.

Some inferences drawn directly from the dimensional universal constants of nature are noted. One of these refers to the existence of a set of ultimate rational units of which all matter and phenomena are constituted. This would place G. N. Lewis's conjecture [1] on the existence of ultimate rational units on a firm foundation by basing it in the most fundamental facts of physics. Also presented are some ontological consequences inferred from these constants, as well as, and independently from, a study of the concept force based on the Mechanics of Gauss-Hertz. These bear on the space-time manifold of physical phenomena, and point up the fact that physics is essentially ontological and that its fundamental aspects are therefore necessarily extra-formal. In a very particular sense this brings back the notion of an aether as the seat and support of all natural phenomena and having an essential physical-ontological property which is ascribed here to point-like impenetrability. Another inference drawn from the dimensional universal constant points to a general connection between certainty and thus determinacy in the domain of the irreducible phenomena in which the constants are entrenched and from which they emerge; and uncertainty in a domain restricted to observables which they determine and with which they are irreversibly connected in the sense of hidden constraints.

The notions of covariance and of correspondence are also considered as particular aspects of the central idea of constancy which is examined in this paper. In so doing, the notion of Universal Correspondence is introduced and compared to the idea of asymptotic correspondence as it is invoked with Correspondence Principles of Physics. Reference is then made to the Universal Entropy Constant found by O. Sackur [2] as well as to the Universal Gravitational Constant, which appears in Newton's law of Universal Gravitation as it pertains to classical mechanical systems, in order to illustrate the notion of Universal Correspondence.

▸ *Constants of nature.* The notion of constancy is primitive, for it appears basic in the function of the mind, in so far as it is necessarily evoked by the mind as it attempts to order and comprehend experience. It is invented by the mind in order to function, because as it contemplates and reasons it must fix attention on things definite, which are therefore in a very broad sense endowed with constancy. The notion of constancy is psychologically linked with the notions of definiteness and regularity and indeed with the notion of order itself. Only with such a preconceived idea can the mind search with confidence and expectation and ultimately discover order in experience, culminating in propositions according to which phenomena become structured in the mind and thus comprehended. The concept of natural law which is invoked in the search for such propositions is indeed preconceived, and is a particular but significant example of the notion of constancy entrenched in our minds. Accordingly, a natural law usually asserts a constant and thus immutable relationship between a set of observables.

In mathematics this concept emerges in its most fundamental and general constructions, namely in the constancy ascribed tacitly to a collection of elements that belong to an ensemble, which in contemporary mathematics are axiomatically defined by well stated and thus definite propositions. I say 'tacitly ascribed' because it is nonsense to formulate definite propositions about indefinite objects. Indeed, it is by axioms that the abstract mathematical objects, the elements of a set, are made definitive and thereby endowed with constancy.

The algebraic and algebraico-geometrical structure of a polynomial is completely determined by its coefficients and the constancy tacitly ascribed to them. This is apparent in the case of algebraic curves and surfaces, obtained in a geometrical representation of algebraic functions by following Descartes in relating the co-ordinates of points in a cartesian frame to numbers assigned to the so-called variables of a function expressed as a polynomial. In so doing, a definite class of algebraic surfaces emerges when a definite set of numerical values is assigned to coefficients of a polynomial. The symmetry and regularity of the geometrical structure so obtained – and their constancy – is implicitly and stringently tied to the assumption that all distances which are determined by the co-ordinates, are based on the same *standard of distance* to which any number may be assigned as a matter of convention. Assigning such a number does not however constitute a condition, which asserts that the standard distance to which it is assigned is either relatively or absolutely constant over the domain for which the polynomial is defined. By relatively constant is meant here that the standard length may change everywhere in the domain of a polynomial relative to some other standard, but uniformly; and consequently, for one restricted to this domain, the geometrical representation of the polynomial cannot reveal such a change in the standard of length on which the numbers assigned to its coefficients are based. By absolutely constant, is meant here that the standard of length is an ultimate unit of which all lengths in the world are constructed, and which therefore cannot within its space-time manifold be referred to another standard for comparison. Of course the idea of an ultimate unit of length of either kind does not have mathematical significance, if one adopts the current view which strives to separate pure mathematics from an ontological basis; that is, from things that actually exist. But if, instead, one regards mathematics and particularly the processes of its invention as an aspect of nature, then it becomes, necessarily meaningful to consider and possibly search for its ontological content. In so doing the question concerning the existence of an ultimate unit of length becomes necessarily a question of fundamental significance. This is particularly the case when mathematics is used to formulate propositions as natural laws, and to apply them for the prediction of phenomena which they condition. Because by so doing we necessarily though not explicitly, ascribe an ontological basis to the mathematical presupposition of a relatively constant standard of length, that is, that the standard of length is relatively constant for the whole domain of the metrical-geometrical manifold. The correspondence and agreement found between mathematics and scientific experience accordingly attests, though indirectly, to the existence in nature of ultimate rational units and their relative constancy.

It follows from these considerations, that, in principle, we cannot decide by scientific and mathematical experience, which is here understood to include the application of mathematics to experience obtained by conventional scientific procedures, whether the ultimate fundamental units are relatively or absolutely constant in nature.

In a certain sense this fact is tantamount to an elementary relativity principle which asserts that the question concerning whether the ultimate natural units inferred from the universal constants are absolutely or relatively constant, cannot in principle be decided within the edifice of scientific and mathematical investigation. There is accordingly, in principle, no way of distinguishing between absolute and relative constancy as defined, since if every experience is restricted to one world, then differences between absolute and relative constancy are not discernible. The distinction between the notions of absolute and relative constancy therefore becomes trivial. This is as an aspect of human experience, essentially equivalent to a fundamental relativity principle which pertains directly to the ultimate natural units determined by the dimensional universal constants. The relativity principles of the Special and General Relativity Theories are therefore considered here to be particular aspects of the relative constancy of the ultimate natural units as determined by the dimensional universal constants.

Moreover, conventional scientific experience does not strictly speaking afford a decisive way to determine and thus decide on whether fundamental and relatively constant units exist in nature, because the information it renders is necessarily approximate. However, the order, regularity and symmetry it nevertheless reveals in the category of approximate information is sharpened with our minds, and subsequently conceptualized and formulated with precision by propositions that belong to the category of exact information in which pure mathematics works and to which the propositions of scientific theory belong [3]. It is in this category of information that metrical mathematics and in particular metrical geometry implicitly evoke the existence (in sense of the abstract) of relatively constant fundamental units. When we use mathematics as a language in science, or if we ascribe to the processes that lead to mathematical discovery an ontological basis, that is, a basis in reality, then this by implication requires, for its justification, that there exist in nature fundamental units which are relatively constant in the sense noted above.*

In other words, if mathematical thought and in particular geometrical thought are considered as aspects of nature, which of course they are, *then* it means that as natural phenomena which are produced in the mind, they themselves tacitly and necessarily evoke the existence in nature of relatively constant units

* These considerations accord with R.C.Tolman's Principle of Similitude. See *Physical Rev.*, August 1914.

of distance, that is, units that are relatively constant in the domain over which a metrical geometry is conceived in the mind, and subsequently formulated. Ontologically speaking, if such units are conceptually identified with, that is, conceptually mapped into, elementary processes that exist in nature, then by relative constancy of such units we mean here, that if they do change, absolutely, then they must change uniformly over the domain of its geometry.

In science, and particularly in theoretical physics, the primitive idea of constancy has been silently used in various contexts to conceive and formulate propositions which have led to far reaching advances in theory that were subsequently supported by discoveries in the laboratory. This idea is essentially philosophical in nature, as it is conceived apriori and compulsively by the mind in order to organize and comprehend what it experiences.

The concept of natural law considered as a definite and thus constant proposition is in itself a prime example of a preconceived idea which is entrenched in the primitive notion of constancy and which dominates almost every theoretical endeavour in science. The notion of constancy is further exercised, extended, and tacitly reiterated by progressively invoking the principle of covariance as it pertains to the formal expression of these laws. The idea of covariance, which is indeed an expression of the idea of constancy in a particular context, has served as a working hypothesis which played a crucial conceptual and philosophical role in the conception and the progressive development of Relativity Theory.

Thus, for example, the requirement that Maxwell's laws of classical electrodynamics be covariant in the set of inertial frames in which Newton's Laws of Classical Mechanics are covariant, imposes connections on the space-time coordinates of events referred to different inertial frames, which accord with the Lorentz Transformation. The application of the principle of covariance, in this restricted sense, essentially invokes apriori the condition that the formal expressions of the laws of classical electrodynamics and of classical mechanics be constant in the same set of frames. This of course is the essence of the principle of special relativity and is the basis for Einstein's construction of the Lorentz Group. The observable implications of this principle are that electrodynamical and mechanical phenomena develop the same way and are in this sense constant with respect to all Galilean frames, and therefore such phenomena cannot be used to distinguish the frames, and thus to identify them absolutely. The principle of General Relativity generalizes to all frames the principle of restricted covariance of the special theory which is restricted to Galilean frames. The observable implications of the Principles of General Relativity, which call for the constancy of the formal expressions of the fundamental laws in all frames, are that phenomena unfold according to these laws equivalently and in this sense constantly with respect to all frames. Such phenomena cannot therefore be used to distinguish the frames or to identify them in an absolute sense : Accordingly ,the principle of general covariance expresses a general condition which pertains to the *constancy* in all reference frames of the mathematical forms that express in them the fundamental laws of physics. We see therefore that this powerful and tested principle expresses an idea which is conceived apriori and which expresses therefore a philosophical point of view about the world. This point of view seems to have been in part inspired by Spinoza, whom Einstein esteemed as a philosopher and whose world view he carried over to physics.

The concept of invariance is fundamental in science and it motivates search for invariants in nature. This concept also calls upon the idea of constancy in a fundamental way, by isolating, and referring to, specific attributes of an observable phenomenon which reveal themselves equivalently, and in this context, constantly in all frames. We attach special significance to invariants, and the reason for doing so is apparently motivated by a philosophical point of view. This point of view ascribes to constancy in nature a fundamental significance, and the grounds for doing this are clearly apriori and thus essentially philosophical. Einstein brought to physics a powerful philosophical mind deeply entrenched and disciplined by an equally strong epistemological attitude. It is this which posed and resolved the questions which others more sophisticated in the ways of mathematics failed to recognize and which therefore they could not of course answer. Einstein is the epitome of a Natural Philosopher, and gave, by his being and work, definition to this term, which reveals its full and ultimate meaning.

Another way in which the idea of constancy emerges in formulating propositions pertaining to natural phenomena is in the formulation and application of correspondence principles. Although a comparatively restricted version of a correspondence principle was explicitly introduced by Bohr, particularly with respect to an interface between quantum and classical mechanics ; Correspondence Principles have also been implicitly used and in various contexts in other developments of theoretical physics, such as in relativity and in the thinking that led De Broglie to formulate the physical principles of wave mechanics.

As already noted, the concept of general covariance plays an incisive role in Relativity Theory. It is linked conceptually, however, with an idea called here Universal Correspondence, for it says that a fundamental law, which is discerned by an observer examining natural phenomena in his particular frame and the formal statement of its proposition, *must correspond* to the law discerned by any other observer examining the *same* phenomena in a different frame. This thinking ascribes apriori uniqueness to an event; it does so in referring to the same event, and the conceptual basis for doing this is ontological, that is, primitively ontological in so far as the event is understood to be something that actually exists.

The progressive development of the theory of relativity can also be envisaged, by first invoking what is called here a restricted principle of correspondence and then extending it apriori to a general or universal principle of correspondence. The restricted principle of correspondence demands, on strictly philosophical grounds, that the set of frames, in which the general laws of classical electrodynamics are covariant, *correspond* to those in which the general laws of classical mechanics are also covariant. As this is sufficient to produce the Lorentz Group, it clearly shows that the *restricted principle of correspondence*, which is here a limited statement of the principle of universal correspondence, is fundamentally linked with constancy, and specifically with the universal constant C, as it, also, serves as a basis for constructing the Lorentz transformations.

In effect, using the correspondence principle in this way is tantamount to demanding apriori, that the connections between electromagnetic phenomena and mechanical phenomena as they are revealed in one inertial frame should *correspond* to their connections in any other inertial frame; and so must also the formal statements of their laws.

The principle of universal correspondence* extends the restricted principle of correspondence to all frames, and this corresponds to the extension of the restricted principle of covariance as adopted in general relativity.

By contrast I shall refer here to Bohr's Correspondence Principle as an asymptotic correspondence principle, because it uses the idea of correspondence in a way that differs essentially from its usage in the contexts previously considered. In Bohr's usage, the idea of correspondence refers to an interface at which a restricted theory, which successfully characterizes a class of natural phenomena, merges with a more comprehensive theory, in which certain aspects of nature are represented which do not appear in the comparatively restricted theory. The comparatively restricted theories, which usually emerge earlier in the history of science, are ordinarily referred to as the classical theories. The principle of asymptotic correspondence requires that the formal statements of the more comprehensive theories correspond, at the interface, to the formal statements

* The essence of this principle is the assertion that what is universal, and thus fundamental in nature, can never be incidental in any natural phenomenon.

which mathematically express the propositions of the classical theory. For this to happen certain terms in the non-classical theories must, according to the principle of asymptotic correspondence, become vanishingly small at the interface.

Although this concept of asymptotic correspondence is endowed with some philosophical content and motivation, and has proven to be very powerful as a working hypothesis which serves to connect by theory natural domains, which appear indiscernable to everyday experience, with those which are accessible; it nevertheless underlines a philosophical point of view which I believe to be essentially untenable; namely, that the so-called classical domains differ fundamentally from the non-classical domains with respect to the ultimate processes that must be considered in order to account fundamentally for the phenomena that transpire in them. A consistent formal aspect and consequence of the application of the principle of asymptotic correspondence, is that the terms which become asymptotically small at the interface between the so-called classical and non-classical domains usually contain or depend upon certain universal constants. This means that certain universal constants that are fundamental for the characterization of non-classical domains are, according to the principle of asymptotic correspondence, irrelevant to what transpires phenomenologically in the classical domain. I believe that this position is philosophically unsound because it is based upon an artificial classification of natural phenomena designed to accommodate the progressive historical development of incomplete theories, and to reconcile comparatively more complete theory with less complete theory. It is not a principle which enunciates a philosophical position about the world, as does for example the Relativity Principles of General Covariance and the Principles of General or Universal Correspondence, introduced here and discussed previously in the context of Relativity Theory. It is instead a working hypothesis which guides the progressive development of incomplete theories and thereby maintains operational contact between theories that refer to phenomena that are remote in the everyday world, that is, to ordinary experience. As noted, it is therefore not a statement about the actual world but rather a statement which is motivated by essentially operational considerations and used for operational convenience. Another and more important reason for deciding here that the principle of asymptotic correspondence is philosophically untenable if it is considered to be a proposition about the actual world, is that its usage implies that certain universal constants are in theory and in fact, inconsequential in the so-called classical domains.

If we regard, as I do here, the existence of the constants of nature as the most fundamental facts which physics has revealed to man about the ultimate processes in nature, then in principle we must expect these ultimate processes to fundamentally and equivalently participate in every aspect of nature and in particular in the so-called classical domains of physics, chemistry and biology. From this it follows that adequate theoretical representations of the so-called classical domains of natural phenomena must in principle include all the Universal Constants of Nature. This then would preclude the elimination of these constants by a procedure based on the principle of asymptotic correspondence. If this philosophical position is in fact valid, it means that the so-called classical theories are in principle incorrect, even in the domains for which they have with satisfactory approximation predicted actual events, insofar as their constructions, and the elementary pictures of nature, which they use for this purpose, are based on incorrect preconceived ideas concerning the fundamental processes in nature. As a manner of speaking we could then say, that the so-called classical theories are based on 'as though it were pictures' which attempt to reconcile constancy and change in nature, rather than on pictures that realistically correspond to the actual processes. These 'as though it were pictures' are evidently very useful, as borne out by the fruits of the classical theories that are built on them, but this of course does not mean that these pictures are true to nature. Indeed the existence of the Universal Constants of Nature can be shown to be incompatible with the 'as though it were pictures' used to formulate the classical theories of physics and thereby fundamentally challenge their validity. The principle of asymptotic correspondence is then a useful working hypothesis for connecting so-called classical theories which are evidently based on incorrect but useful elementary pictures, with relatively more complete theories, in which modelling as a way of picturing fundamental processes in a space-time manifold is relinquished, or at least de-emphasized. Evidently, it is because the classical theories are constructed on the basis of incorrect models which attempt to reconcile constancy and process in nature, that accounts for the absence of certain universal constants in the formulation of their propositions.

Whereas these incorrect models may be adequate for constructing a theory for the classical domain which pertains strictly to mechanical phenomena, they may prove inadequate to account for Chemical-Mechanical phenomena which are so fundamental in the life processes, and which in a sense may be considered to belong to the classical macroscopic domain in nature ; the basic question here being : how are so-called classical mechanical forces made, chemically, in biological systems ?

In this regard it is relevant to draw attention here to the work of O. Sackur [2], who showed that the constant in the thermodynamic equation for the entropy of a perfect monoatomic gas could be separated into two terms, one depending upon the atomic weight of the gas, and the other being a constant independent of the nature of the gas and consequently a universal constant of all monoatomic gases. Sackur attempted to calculate the value of this constant from the theory of quanta, but obtained values which are not consistent with experimental data. G.N. Lewis and Adams [4] then calculated the Sackur Universal Entropy Constant for Monoatomic Gases, by using ultimate rational units introduced by them in an exciting paper entitled 'Notes on Quantum Theory' published in 1914 in the *Physical Review*. Using their calculated value of this Universal Entropy constant Lewis and Adams calculated the entropy of four monoatomic gases at 25°C and one atmosphere and obtained reasonable agreement.

Although the theoretical basis presented by Lewis and Adams for the existence of ultimate rational units appears unsound in the judgment of some of Lewis's most accomplished followers, I am convinced, however, that G. N. Lewis's belief in the existence of natural irreducible units can be established on the basis of the dimensional universal constants of physics, according to reasoning and a procedure which Lewis apparently did not consider. In this regard it appears relevant to report that work based on the dimensional constants of nature led me some years ago to infer from a single and apparently sound hypothesis, that these constants do indeed imply the existence of fundamental natural units. This may in effect establish G.N. Lewis's belief in and use of Ultimate Rational Units on what may be considered the most solid foundation science can render at this time, namely, the Universal Constants of Physics.

It seems particularly important to draw attention here to the fact that all terms which appear in the expression for the Universal Entropy Constant obtained by Lewis and Adams, are individually Universal Constants. These consist of the velocity of light c, the molecular gas constant $k=R/N$, the molecules in a mol N, and the electron charge e. It is interesting to observe that the work of Sackur and Lewis, which led Lewis to the calculation of a Universal Entropy Constant in terms of certain Universal Constants, is based on and is therefore restricted to systems characterized by the Universal Perfect Gas Law. This may give and probably has given the impression that this result has limited significance. However, some inferences drawn directly from an examination of the Universal Dimensional Constants of Physics have shown me that the perfect gas may indeed serve as a realistic model for characterizing a most fundamental aspect of nature, which the existence of these constants directly imply. In this regard, consideration is given to the fact that what characterizes the monoatomic gas is not the indivisibility of its constituents but rather its adherence to the law $C_p = 5/2R$, and it may therefore be assumed that any gas, in the ranges in which this law is obeyed, will follow the expression for the entropy calculated by Sackur, and which contains the Universal Entropy Constant evaluated by Lewis by using his 'Theory of Ultimate Rational Units'.

The above considerations give strong and somewhat concrete support to the thesis presented here earlier, according to which the so-called classical as well as non-classical phenomena are equally entrenched in and supported by the same irreducible natural processes of which the Universal Constants are but manifestations. This is an aspect of what is called here the Principle of Universal Correspondence. The reason for saying this, is, simply, that the entropy of a monoatomic gas has a fundamental significance in the macroscopic and so-called classical domain; and since, according to Lewis, Adams, and Randall, its absolute value depends directly on a set of Universal Constants, it is clear that they must impinge fundamentally on phenomena now arbitrarily ascribed to the classical regime. Clearly, such fundamental intervention of a set of Universal Constants in phenomena ascribed to the classical regime cannot be revealed in such cases where the principle of asymptotic correspondence is used, for its application would annihilate their contribution to the formal statements of the propositions which are considered appropriate to the classical regime. The same kind of reasoning pertains to the presence of the Universal Gravitational Constant in Newton's Law of Universal Gravitation formulated in the edifice of classical mechanics.

Another context in which the central idea of constancy is underlined in the fundamental propositions in Science, is in laws which call for conservation with respect to particular attributes ascribed to states of matter. The laws which are considered most fundamental in science are indeed of this kind. The requirement that *an aspect of nature* be subject to a conservation law and is thus conserved, is of course equivalent to the condition that it be and remain *constant*.

Among the physical theories, the Special Theory of Relativity is unique, in so far as it formulates, in terms of spatial-temporal pictures and observables by which we describe and record our experience, the proposition that the speed of light is a Universal Constant, that is, *that the velocity of light is independent of the motion of the light's source*. Einstein showed that only this single and simple axiom in electrodynamics need be assumed and formulated in order to establish the covariance of all laws of nature under the Lorentz Transformation. This means that the Lorentz transformation and all of its remarkable and largely incomprehensible consequences can be ascribed to the existence of the Dimensional Universal

Constant c, and to a proposition which formulates and thus explicates this fact in terms of space-time observables that are referred to Galilean frames.

The Special Theory of Relativity is apparently the only theory which formulates a proposition which explicitly invokes the constancy of a Dimensional Universal Constant. Quantum Mechanics does not do this with respect to the Quantum of action, and therefore does not explicitly formulate the proposition that the quantum is a Universal Constant. These considerations lead to a definite and evidently meaningful criterion for distinguishing between ultimate and non-ultimate physical theories. The former are directly based upon and *explicitly invoke* the constancy of at least one Universal Constant of Nature, whereas the latter do not. From this it follows that the Special Theory of Relativity is an ultimate theory – and this may account for its powerful and universal consequences which have been experimentally verified – whereas quantum mechanics is not an ultimate theory. Although the universal action constant does appear as a constant of proportionality in the propositions of quantum mechanics, these propositions do not however formulate the universal constancy of the quantum constant.

Some years ago I was struck by the realization that the Constants of Nature, as they are revealed to us in physics, are the most fundamental manifestations of the ultimate processes in nature and of their attendant laws. This motivated a search for ways of inferring, from these constants, conditions and constraints on the nature of the physical space-time manifold which their existence would necessitate and thus impose, as well as on the ultimate processes embedded in this manifold, which are presumably indirectly revealed by the Universal Constants. In so doing inferences have been drawn which gave reason to re-focus attention on the concept of a Neo-Aether, envisaged here as a continuously extended substance, whose essential ontological property is point-like impenetrability. With one basic exception, this idea seems in part to accord with a suggestion Einstein made in 1920 in Leyden, concerning the extension of the notion of an Aether. In this he explicitly stated that the Aether should no longer be regarded as a substance but simply as fields, etc., or the totality of those physical quantities which are to be associated with matter-free space.

The present concept of a Neo-Aether emerging from the Dimensional Universal Constants, finds support in the sense of a correspondence principle in a study [3] concerned with the foundations of classical mechanics and more specifically with attempts made by Gauss [5] and Hertz [6] to rationalize the concept force by placing it upon a geometrical foundation. This study shows that the mechanics of Hertz, which brings Gauss's programme to fruition, is in fact an incomplete theory because it does not refer the formal geometrical conditions used to geometrize force to an ontological basis, that is to a basis in experience. This of course is essential to the construction of a map which can relate their formally stated propositions and their formally derived consequences to experience ; otherwise the geometrization of forces, which they carry out, is but an exercise with symbols on paper. The only way which I could find to give ontological support and thus a basis in reality to the geometrical restrictions used by Gauss and Hertz to establish force on a geometrical foundation, is by introducing the idea that these geometrical restrictions must be ascribed to the pointlike impenetrability of matter. It is accordingly becoming increasingly clear that impenetrability may prove to be the most fundamental ontological property of substance, that is, the ultimate property by which substance evokes its existence and thus identifies itself in space time.

I will therefore attempt to shed further light on what appears here to be a fundamental connection between the production of forces in nature, the evolution of geometrical forms in its space-time manifold, and impenetrability as the fundamental ontological property of this manifold. It is with respect to such a manifold that the notion of a Neo-Aether was noted above. For this purpose it is helpful to briefly consider some of the crucial ideas underlying the classical mechanics of Gauss and Hertz, and explain in further detail why their theory, which was motivated by a search for a geometrical and thus rational foundation for force, is scientifically speaking incomplete.* The incompleteness of their theory, which otherwise is endowed with far-reaching and unexplored implications, can be ascribed to the fact, that the geometrical restrictions on the classical space-time manifold, which they formally impose, were not identified with or justified in terms of actual experience. This is another way of saying that the incompleteness of their theory can be attributed to the fact that it is devoid of an ontological basis, which reflection shows to be essential in the conception of any process that can produce a force ; because force itself is necessarily ontological, in so far as it is a manifestation of things that actually exist and, is therefore inextricably linked with experience.

Gauss formulated a general and fundamental principle of mechanics by ascribing to force a purely geometrical and thus formal basis. This was motivated by an attempt to make precise the concept force, by entrenching it in the primitives of geometry. Gauss's motivation apparently stemmed from the observation that Newton's fundamental proposition may be considered to be a statement which

* These considerations also have bearing on the formal geometrization of the space-time manifold followed by the General Theory of Relativity.

defines force in terms of motion, rather than a principle in which the concept force is regarded as primitive, and which, I believe, corresponds to Newton's own interpretation. This conclusion is based on the fact that Newton ascribed a priori, to the symbol $\overline{F}$ (and therefore to what it represents in nature) an unrestricted covariance, in the sense that in the general case the force $\overline{F}$ so to speak belongs to the body on which it is impressed, and is therefore the same in any and all frames of reference to which the motion may be referred. In particular, therefore, it is the same force for non-inertial as well as for inertial frames. The force that the symbol $\overline{F}$ represents in Newton's fundamental proposition is the sum of all the mechanical interactions a material particle experiences and, thus, senses in the universe; and as such, represents an intrinsic connection between the particle and the universe, as this connection is independent of any particular frame in which its consequences, say in motion, may be examined.

The above way of thinking about force and interpreting the condition $\overline{F}=\overline{F}'$ prescribed a priori by Newton and supported by experiment, is apparently connected with Mach's Principle, because the same interpretation was given by Mach to the inertial mass of a body. From this it follows that the force $\overline{F}$ to which Newton assigned a priori unlimited covariance, is in general more fundamental than is the acceleration of a particle, which limits the covariance of Newton's fundamental proposition to inertial frames. It is interesting to note that the inertial mass in $\overline{F}=m\overline{a}$, which according to Mach is also a manifestation of a connection between the particle and all matter in the Universe, enjoys the same unrestricted covariance as does $\overline{F}$, and is met naturally in a formal way because it is a scalar. From these considerations based on classical mechanics we are led to the conclusion that the concept force is indeed more fundamental than is the idea of a particle in motion, which may be thought of as a very special and restricted manifestation of force within realms of classical mechanics. If this is correct, then it is naive to regard Newton's fundamental proposition as a statement of definition of force.

Even though what motivated Gauss's search for a new formulation of the principles of classical mechanics appears to be invalid, his variational principle of mechanics has imparted to classical mechanics new dimensions and areas of inquiry; particularly, as the procedures he uses to geometrize force do not necessarily conflict with the unrestricted covariance which Newton ascribed a priori to force.

Gauss's Principle is a differential variational principle that consists of one statement which refers to and conditions a global scalar measure of a mechanical system, consisting of an arbitrary number of discrete bodies subject to impressed forces. To my knowledge, it and Hertz's variational principle are the only strictly minimum principles of classical mechanics. Gauss's Principle does indeed therefore *directly* express and communicate information which pertains to the macromechanical processes that are produced in many-body systems.

As noted, Gauss formulated his new and fundamental variational principle of mechanics by ascribing to force a purely geometrical and thus formal basis; but he did retain, in his formulation, forces impressed on particles from sources outside of the configuration. Hertz extended and completed Gauss's work by ascribing to all forces in nature a geometrical basis. The reduction of all forces to geometry, as effected by Hertz, is strictly a formal process whose scientific justification now appears to be contingent upon extra-formal, i.e. ontological considerations. This means that the very basis in nature for the geometrization of force by Gauss and Hertz may be extra-mathematical; that is, in principle not amenable to formal representation. This question which concerns the basis in nature and thus the justification in experience for the geometrical constraints used by Hertz to geometrize all forces is most fundamental because it must be considered, and at least conceptually resolved, in order to relate the formally stated propositions of Hertz and their formally obtained consequences, to experience. Otherwise Hertz's Mechanics remains, scientifically speaking, a sterile, i.e. a dead document. The fact that this question had apparently not been recognized and dealt with sooner, may account for the fact that the potential of Hertz's profound conceptual edifice has remained dormant for so long.

A crucial idea, which in my opinion is the key to the resolution of this basic question, is the fundamental connection that exists between the impenetrability of matter and the production in nature of stringent geometrical constraints. This idea, which was introduced and used in Refs. 7–12, merges geometry with substance, and underlines the existence in nature of a substantial basis for its geometry which is here attributed to the impenetrability of matter. In other words, impenetrability is here considered as the ontological and thus physical basis and support for the geometrical constraints by which Hertz formally geometrized force. In this view, the geometry of nature is embedded, that is sculptured, in matter by virtue of its impenetrability which is also the basis for the production of templates in nature, which are of course important in current biological thought. It appears therefore according to this thinking, that ultimate processes involved in natural phenomena do not only concern the geometrization of matter but equally if not even more fundamentally, the materialization of geometry.

The notion of impenetrability is essentially ontological and it therefore does not have a formal representation. From the above considerations, it appears to be the crucial concept by which the geometrical ideas formally developed by Hertz can be related to experience and, in this way, bring the mechanics of Hertz, at least conceptually to *completion* as a scientific theory. By his important work, Hertz has forced us to think in a way which has revealed what may be a universal connection between force and impenetrability ; the geometry of nature being but a manifestation of this universal connection.

A necessary condition that a scientific theory be complete, is that it give consideration to its ontological basis, that is, to the basis in fact, in experience and thus in nature (which support its fundamental propositions) ; and that it specify, at least conceptually, the rules and operations by which to relate its formal propositions and their formal consequences, to experience. By a scientific theory is meant here, a theory consisting of propositions that are presumed to pertain to the real world, and which can lead to the conception of experimental arrangements and experiments, in which and by which they can be tested.

From these considerations it follows that the mechanics of Hertz, which affords greater range and depth in formal expression, and which is more general in its formal structure than is the mechanics of Newton, is nevertheless an incomplete scientific theory. The reason is that it lacks an ontological foundation, as explained above. In its original construction, it is strictly a formal theory, consisting of symbols and specified operations upon them which are devoid of a map that imparts to them experimental content. Hertz was of course deeply concerned with experimentation, and therefore questions concerning the verity of his mechanical edifice and the propositions he formulated within it, were of paramount importance to him. In the regard, however, he appealed to the fact that Newton's fundamental proposition $\overline{F}=m\overline{a}$, can be derived from his proposition, by associating certain combinations of symbols obtained from his theory with the symbol $\overline{F}$, representing the concept force in Newton's Mechanics.

In this way we see that Hertz's mechanics, which was motivated by a search for a rational basis for force in geometry, and thereby for its elimination in the propositions of mechanics must nevertheless, though indirectly, refer to force by way of Newtonian Mechanics, in order to relate its own propositions and their consequences to mechanical experience. According to this procedure, Newton's mechanics furnishes a bridge and thus a map which relates Hertz's formally stated propositions to scientific experimentation. Since Hertz's mechanics does not provide such a map within its own edifice, it is therefore incomplete as a scientific theory. However, the fact that Hertz's mechanics yields formally the proposition of Newton, attests to the fundamental significance of the formal representation of the geometrical constraints it uses to geometrize force as an aspect of nature; thereby giving indirect but strong support to the concept that the impenetrability of substance is the physical-ontological basis of forces in nature.

SUMMARY

In summarizing, it may be of interest to point out some inferences and guide lines drawn from the study based on the Universal Constants of Nature, previously noted. This I hope will make clear that the fundamental questions that emerge from an examination of these constants are to a large extent conceptual and thus philosophical in nature and have profound scientific implications. The fact that they are tied to these fundamental constants entrenches them in the most fundamental facts and results, which, in my opinion, science has produced to date.

The inferences drawn from the dimensional constants of nature relate to a domain in which the irreducible natural phenomena from which these constants emerge are entrenched. It is to this domain and to these irreducible phenomena that the immutable and thus deterministic relationships tacitly implied by the very concept of a natural law, are ascribed here. In so far as the inferences drawn are here based upon the most fundamental and evidently most universal facts which science has so far produced, they do not fall within the realm of metaphysical explanation as described by P. Duhem,* according to which the fundamental propositions of physics are to be derived from notions introduced a priori and which therefore do not afford a common working medium for arriving at a consensus concerning them. The physical foundations of the propositions of physics are here examined by starting instead with its most universal facts, and deepening our knowledge and understanding of physics by searching synthetically for a conception of the space-time manifold which can accommodate these elementary facts.

This is somewhat analogous to the techniques used in metamathematics in which the foundations of mathematics are examined by using the most universally accepted problems of mathematics itself, thus allowing for consensus.

1. I find that the Ultimate Constants of Nature are the constants which, so to speak, have physical dimensions. From this it follows that propositions which evoke their constancy, formally or otherwise, cannot be given a numerical representation. If by hypothesis we identify the dimensional constants with the ultimate

* P. Duhen, *The Aim and Structure of Physical Theory*, Princeton University Press.

processes in nature, that is, if we interpret them as manifestations of these ultimate processes within the domain we now call physics, then it follows that their laws, as determined by the existence of these constants, are necessarily qualitative. An example of this is the fundamental proposition of the special theory of relativity which invokes the constancy of the speed of light. I realize that this does not accord with an accepted view which holds that the ultimate constants are those which can unambiguously be represented by pure numbers. The case for this position is strong and it took me some years to break away from it. It is apparently motivated by the preconceived idea that what is ultimate in science is necessarily quantitative, which of course disagrees fundamentally with the above conclusion, which says that what is ultimate is necessarily qualitative.

2. The existence of the Dimensional Universal Constants imply that the ultimate processes with which they are identified in nature are in principle inaccessible, that is, the processes from which they emerge are at least relatively constant and thus immutable with respect to any and all changes that emerge in the so-called observable domain; and in particular therefore in the domain of human experience and human action. The term 'observable domain' as used here, corresponds to Kant's reference to nature as 'the complex of objects of external sense'. It is accordingly convenient to refer here to the domain of the ultimate processes, as the inaccessible domain in nature; for were it accessible then the ultimate processes which are – according to the dimensional constants – immutable, would then be amenable to change, which would produce a contradiction.

3. If we enlarge our view of the Dimensional Constants, and think of them as the constants of nature, even though they in fact emerged in the realm we now call physics, then it follows that they are indeed the Universal Constants of the life process as well, and it is to them that we must ultimately turn for a scientific understanding of this process.

4. The obvious connection between the idea of constancy, and stability as an aspect of immutability, makes it clear that the stability of states of matter which is conventionally and specifically ascribed to the quantum constant, and for good reason, has a deeper basis. It is the existence of a *set* of dimensional universal constants in nature, from which constancy and stability in nature apparently comes, rather than from the particular constant known as the quantum. In other words, the process of quantitization appears to come fundamentally from and to be determined by the fact that the Dimensional Universal Constants imply the existence of a set of relatively constant units, and in this sense ultimate units, which have their ontological basis in matter.

5. The concept of natural law necessarily evokes the notion of constancy, in so far as it refers to definite and thus constant propositions about the world. In physical theory, the laws refer to observables by which phenomena as processes are described and measured. The observables which these laws condition and thus constrain are consequently variable in the general case, and the processes in nature by which they are constrained according to the laws are not described by the observables themselves. It follows directly therefore that the observables constrained by a law are not strictly determined by the phenomena which they describe. The inherently statistical nature of observable natural phenomena is accordingly considered here to be a consequence of this inherent indeterminacy which necessarily exists in the restricted domain of the observables, as they are strictly determined by phenomena which they do not describe; that is, by (extra-observable) phenomena which do not belong to the observable domain. This is particularly the case if the constant relationship between observables invoked by a law, is determined by ultimate processes from which the constants of nature emerge. Examples of such laws are the propositions of quantum mechanics in the so-called non-classical domain, and Newton's Universal Law of Gravitation in the domain of classical mechanics.

6. A formidable challenge to our minds is to find a way of reconciling ontologically two primitive notions that confront the mind compulsively, which therefore in the spirit of Kant I take to mean, have essential verity, namely: constancy and change; and to search for pictures (in the sense of Wittgenstein) entrenched in the space-time manifold by which such a reconciliation can be conceptually effected. In so doing we may find a way of accommodating the dialectic process as a fundamental aspect of nature; and possibly discovering how the ultimate processes follow it in the sense of an extended and deepened ontological conception of complimentarity. This may lead to a conception of evolution as an ultimate process in nature.

7. The accommodation of the constancy and thus inaccessibility of the ultimate processes in nature as implied by the existence of the Dimensional Universal Constants, evokes the conception of a Neo-Aether whose essential ontological attribute is pointlike impenetrability.

▸ *Concerning natural law and biological theory*. The ultimate facts reflected in the existence of the Constants of Nature do not imply atomism. In fact, this point is missed when they are referred to (as they usually are) as the *Atomic Constants*. The Constants of Nature in fact deny atomism, that is, that the elements of nature from which they emanate are *'particulate'*, that is, made of matter with which they are constantly identified and to which they therefore bear a constant (fixed) and consequently static relation. On the contrary, I find that the Universal Constants demand that the elements of nature be endowed with a process of regeneration by which the constancy of their space-time structures is sustained and constantly regenerated in matter which belongs to a locally impenetrable material substratum; that in this way constancy and process are inextricably and inseparably linked in and by the elements of nature from which all phenomena and forces in nature ultimately emerge and with which they must be reconciled. The elements so conceived and the intrinsic connection between constancy and process with which their existence is necessarily endowed and on which it depends, indeed determine finite ultimate natural units, which include an interval of natural time, that is, duration. By this reasoning all natural phenomena derive impetus, thrust, and organization from the inherently dynamical elements of nature, and are spatially and temporally co-ordinated by the condition that they be congruently made of the ultimate natural units with which they are endowed. This I believe is the basis of space-time structure, and in particular of temporal organization and critical co-ordination and performance of biological material in which the ultimate processes in nature as reflected in the constants of nature literally come before our naked eye.

In a sense (which I hope to present in another paper) this emergence of the ultimate and universal processes in biological material is, I believe, an aspect of a general process of evolution to which biological evolution in its presently more restricted sense belongs. By this thinking, all natural phenomena are particular aspects of this general evolutionary process, which is based on a connection between two domains in nature. One I call the Carnot domain (the domain of ordinary sense experience) to which science in its narrow sense is by convention committed. The other is the domain of the Universal Constants of Nature and the ultimate processes (which are immune to Carnot's principle) from which they emerge. It is the second domain which gives nature its thrust, its organization and its organism, and which drives and progressively evolves into the first by a general process of evolution. According to this concept the reason the history of nature does not comply with Carnot's principle, is that the domain of the Universal Constants constantly and irreversibly impresses (by impressing *active* stringent constraints) on the space-time manifold structure and thus non-uniformity, which averts the evolution of the thermal death (the totally uniform heat bath) implied by Carnot's Principle which applies strictly only to the first domain.

We are of course unfolding here a conception of nature that corresponds in some features with Leibniz's thinking. Accordingly, the domain of the Universals bear, some correspondence with Leibniz's Concept of Monads which remained highly metaphysical and vague in his work. Here, however, the conception of the domain of the Universals is based on the most fundamental and universal facts that science by its severely objective methodology has produced, within the edifice of physical theory and experience. These facts, of course, concern the indirect identification of the Universal Constants of Physics. This conception of nature does indeed challenge presuppositions which lead to and adhere to descriptions of nature that underline the relationship between observable (conceived as an object) and an observer externally related to the object. Instead, we see here that a deeper grasp of nature, and that ways of evolving new and more direct ways of learning about nature by experimentation, may come from the realization that the essential relations in nature are internal, and cannot be grasped in terms of presuppositions that stress the external relationship between man the observer and nature the observable.

These ideas that bring forth here new presuppositions call for a direct interplay between conceptualization in all ramifications and experimentation, and the realization that this is the epitome of theory in Science, and of biological theory in particular; that the ultimate laws of nature are tied to the Constants of Nature, and consist of propositions (if they can be explicated) which refer directly to these constants; that, in particular, these are also the laws of Biology. Conceptualization in all of its ramifications is not necessarily amenable to mathematical formulation, particularly as it pertains to natural phenomena. In a previous paper I explained the inherently passive, a-ontological and the a-phenomenological aspect of mathematics, both of which are of course essential in thinking about nature in which ideas concerning sensation, process, and actual experience are crucial.

The direct correspondence between thinking about nature, and experimentation, must be recognized and emphasized as the ultimate instrument of biological theory. The fact that process is an essential and a particularly prominent aspect in the description of biological material, accounts I believe for the fact that so much of the truly outstanding thinking based on these descriptions, and the ideas and principles which they have evoked, have not been embraced in mathematical language, which is essentially devoid of process. This does not mean of course that mathematics has not much to contribute to biological theory, but it does mean that it is an incomplete instrument which must be complemented by extra-formal thinking.

Paul Lieber

The thinking which refers directly to aspects of nature which are inherently extra-formal, must and can be related directly to the real world, by testing hypothesis, conjecture, inference, and questions which it produces by the conception and design of crucial experiments. The information content of any experiment, and particularly a crucial experiment, depends on the hypothesis, the conjecture, the inference, and the questions which led to its conception. It is this which distinguishes information from data.

With this background, some ten years ago I conceived a crucial experiment with biological materials which concerns the nature of adaptation conceived as a process in the space-time manifold. During this period I have examined results of other experiments that appear from the standpoint of their experimental arrangements related to the experiment I have in mind. This was done in order to determine whether their results deny, are consistent with, or support the thinking and hypotheses which led me to the conception of this experiment. The highly important work of C. H. Waddington * which led him to the penetrating concept of genetic assimilation of an acquired phenotypic character, appears to be particularly related to this experiment, and is consistent with the hypothesis by which it was conceived. Equally supporting are recent results concerning repair of damage of Bacterial R N A.

▶ *On the resolution of a fundamental question posed in a dispute between Poincaré and Russell.* † In modern expositions of the axioms of geometry, conditions are given in which the relation of congruence between segments is to satisfy. By so doing it is supposed that we have a complete theory of points, straight lines, planes, and order of points on planes – and thus a complete theory of non-metrical geometry. When we then consider congruence and the set of conditions – or axioms – which this relation satisfies, we find that one can prove that these alternative relations satisfy these conditions equally well and that there is nothing intrinsic in the axioms of non-metrical geometry which gives preference to a particular relation of congruence. That is, different metrical geometries are equivalent as far as the axioms of intrinsic (non-metrical) geometry are concerned.

In Poincaré's view, the choice made among these geometries for the description of nature is guided purely by convention, and that the effect of a change of choice would be simply to alter the expression of natural law. By this view, there is nothing inherent in nature that gives a peculiar role to a particular one of these congruence relations, and the choice is made on the grounds of intellectual

* C.H.Waddington, *Evolution*, 7, 118–26, June 1953.

† A.N.Whitehead, *Concept of Nature*, Cambridge Press.

convenience rather than on the basis of natural fact. Poincaré asked for any factor in nature which gives pre-eminence to a particular congruence relation, such as the one adopted by mankind. Russell pointed out that according to Poincaré's principles there was nothing in nature to determine whether the earth is larger or smaller than a given ball. By this example Russell in effect pointed out that apart from minor inaccuracies a determinate congruence relation is among the factors in nature which our senses and experience posit for us. Russell's position was further strengthened by the assertion 'As a fact of observation, we do find it, and *what is more, agree in finding the same congruence relation*.' In response, Poincaré *asks for information as to the factor in nature which might lead any particular congruence relation to play a unique role among the factors presented in sense awareness*.

Whitehead, who recognized the importance of this controversy, realized that its resolution is of utmost importance, and furthermore that it implies that the fundamental factor called for by Poincaré indeed remained to be found. Poincaré's reasoning and position are incisive, as is Russell's observation, which does not in fact answer Poincaré. The answer resides, I believe, in the Constants of Nature which are the ultimate facts of nature which embody the factor called for by Poincaré. Furthermore, Russell's observation is, I believe, an aspect of a general process of evolution through which the constants emerge in thought, thereby giving a basis for agreement.

CONCLUSION

In conclusion, a number of concisely stated inferences based on a study of the Universal Constants are summarized below:

1. That the Dimensional Universal Constants are in a significant sense the most fundamental universal constants in nature. In this respect I cannot agree with Eddington,* who a priori ascribed the most fundamental role to the dimensionless universal constants.
2. That the irreducible laws of nature are propositions which evoke and formulate the universal constancy of the dimensional universal constants.
3. That these propositions, that is, what they say, cannot be expressed numerically; and in this meaning they are necessarily qualitative.
4. That the dimensional universal constants determine naturally rational criteria for an ultimate scientific theory.
5. That among the existing theories, the only theory which satisfies these

* A. Eddington, *Fundamental Theory*, Cambridge Press.

criteria is the Special Theory of Relativity. This inference is supported by the fact that the special theory, or what is formally equivalent to it, namely the Lorentz group, is now being used as a fundamental restriction and guide line on any acceptable development in relativistic-quantum mechanics, in other words, quantum-electrodynamics. I speak here of the stringent requirement that all current developments in quantum-electrodynamics are a priori required to be Lorentz invariant.

6. That quantum-mechanics is not an ultimate theory.

7. The conjecture, that the universal constants of physics are indeed the universal constants of nature and therefore the universals of the life process as well. They have simply emerged within the edifice of what is now by convention called physics.

8. That biology, chemistry, and physics merge on the level of the dimensional universal constants and that fundamental propositions pertaining to the biological processes merge with those of chemistry and physics only when they each satisfy the criteria of an ultimate theory. That the ultimate propositions of theoretical biology are, therefore, not reducible to those propositions of physics and chemistry that are not ultimate.

9. It follows that the ultimate propositions of biology are not reducible to those of physics, as physics is known today, with the exception of the special theory of relativity. When the ultimate propositions of biology will be reducible to those of physics, they then will become one.

10. The dimensional constants of nature deny the existence of elementary particles, conceived as bodies which convect, that is transport, their matter as they move; that is, carry the material of which they are made with them as they change position.

11. The dimensional universal constants imply the existence of a set of irreducible units in nature; this inference gives support to G. N. Lewis's conjecture and search for such units and on which he based his calculation of the universal entropy constant of a perfect gas, following the work of O. Sackur.

12. That the dimensional universal constants necessarily imply the existence of a non-observable domain in nature which is the seat of the universal processes, and therefore, the existence of an interface between the observables, to which scientific experience has committed and thus limited itself as a matter of convention, and the domain of the non-observables, which cannot be conceptualized in terms of pictures based on subject-object relationships, i.e. the observer and observable pictures, in terms of which the domain of the

observables is now conceptualized. The hidden variables of mechanics (ignorable co-ordinates) and those referred to (in thermodynamics) by Caratheodory evidently belong to this non-observable domain.
13. The dimensional constants, that is their existence, necessarily evoke a conception of a neo-aether, that is, a physico-geometrical manifold, whose essential ontological property is point-like impenetrability.
14. That in the sense of a correspondence principle this inference is supported by a study in classical mechanics concerned with establishing a geometrical-ontological basis for force.
15. That the Michelson-Morely experiment is compatible with the existence of a neo-aether everywhere endowed with this essential ontological property.
16. That the non-observable domain suggests that the life process is controlled and executed on a sub-atomic level.
17. This has led to conjectures and the conception of crucial experiments with biological materials by which to test them. It is planned to carry out these experiments in the Donner Laboratory at the University of California at Berkeley.
18. That the manifold of the physical space is endowed with an essential ontological property which is in principle extra-mathematical and which relates to impenetrability.
19. That this property appears to be fundamentally linked with symmetry properties which have emerged in contemporary nuclear physics and which are fundamentally connected with mirror-symmetry, that is with right and left hand symmetry.
20. Also inferences have been drawn from the universal constants which have pointed up the need to form a new conception of the ultimate categories of experience, with respect to those considered by Kant in his Critique of Reason. This new conception must reconcile in a fundamental way, immutability, that is, constancy, with process, and must endow conceptually the space time manifold with ontological properties by which such a reconciliation can be effected.
21. That the existence of the non-observable domain in which the immutable processes in nature and their fundamental laws are seated, necessarily implies that laws restricted to the observable domain, that is, laws that refer strictly to observables, are necessarily statistical.
22. That quantization is an aspect of universal constancy, rather than a consequence of a particular universal constant.

'Error comes from exclusion.'

– Pascal

References

1. G. N. Lewis *Phys. Rev. 18* (1921) 121.

2. O. Sackur *Ann. Physik 36* (1911) 598; *40* (1913) 67.

3. Paul Lieber *Categories of Information.* Reiner Anniversary Volume. (Pergamon Press, to be published).

4. G. N. Lewis and E. Q. Adams *Phys. Rev. 3* (1914) 92.

5. Carl F. Gauss *Creele's Journal f. Math 4,* (1829) 232. Also appears in *Werke 5,* 23.

6. Heinrich Hertz *Principles of Mechanics,* Collected Works, vol. III (Macmillan: New York 1896).

7. Paul Lieber and K. S. Wan. *Proc.* IX Int. Congress of Mechanics (1957)

8. Paul Lieber and K. S. Wan. Air Force Office of Scientific Research Report TN 57–479, AD 136, 471 (September 1957).

9. Paul Lieber and Arthur Farmer *Trans. American Geophysical Union, 39,* no. 2 (April 1958).

10. Paul Lieber and K. S. Wan *Trans. American Geophysical Union,* vol. 43, no. 4 (December 1962) 444.

11. Paul Lieber. Volume of the Proceedings of the Symposium on Second-Order Effects in Elasticity, Plasticity, and Fluid Dynamics, sponsored by the International Union of Theoretical and Applied Mechanics (April 1962).

12. Paul Lieber. Institute of Engineering Research, University of California, Berkeley, NOnr-222(87), no. MD-63-8 (April 1963).

'Boxes' as a model of pattern-formation

Donald Michie and R.A.Chambers
University of Edinburgh

▸ *The French flag.* In discussion of morphogenetic regulation, Wolpert has used the metaphor of the French flag. How could we assemble a display in the form of the French tricolor by appropriately organizing a company of men drawn up in rank and file formation? If each man carries a red, a white and a blue flag, from which he can select one to hold up to view, what rules of next-neighbour interaction would assist their leader to generate from the individual displays the desired overall pattern?

Some local rules, one feels, clearly justify themselves by simplifying the required global instructions. For example, a rule which says 'If the man in front of you and the man behind you agree, then agree with them' not only has a general error-correcting effect but it also ensures that the intended vertical boundaries between the three coloured areas will in fact be vertical and also straight. Compare the 'firing squad' problem where a chain of soldiers, each able to exchange information with his two neighbours, must 'come to an agreement' and fire their guns simultaneously (*see* [1]).

We do not propose to speak further about the French flag. Yet almost everything we shall say can be interpreted as being about the 'French flag problem'. In fact in the experiment which we shall describe we have been working on the French flag without knowing it.

▸ *Central and local organization.* We interpret the essence of the problem to be as follows: a *pattern* is to be created through the interplay, in successive instants of time, of a set of global instructions and a set of local instructions with which an array of automata are to be severally equipped under this central command. By a global instruction we mean an instruction uttered by the leader during the pattern-building process which an individual automaton can obey without reference to its neighbours. As for local instructions, in the simplest case the automata are identically equipped, and this condition characterizes the example which I shall describe. But this is not a necessary restriction. The essential question is 'How far, if at all, might considerations of economy and efficiency suggest a decentralization of command, as against complete specification by the central authority'?

The relevance of this question to theories of morphogenesis is obvious. The value of the example which we shall present, if it has a value, lies precisely in the fact that the work was done in a context far removed from morphogenetic, or indeed from biological, topics. The attempt to solve a difficult problem in adaptive control led us to devise a system of local automata interacting with a global command which presents certain striking parallels to the manner in which some adaptive problems have been solved in Nature.

Our adaptive control problem is of the 'black box' variety. Under the rules of the game a system must be designed which will control some physical apparatus, the properties of which are at first more or less unknown. Further, the system must be self-designed, utilizing its trials and its failures as the only source of information from which to organize itself into an appropriate stimulus-response pattern, meaning by this a pattern giving effective control over the apparatus. The computer program which was written to perform this task thus exemplifies a 'learning', or 'evolutionary', or 'adaptive', or 'self-organizing' system. Since we suppose a definite cost to be attached to each failure during the pattern-building process, the design objective is a system which can attain a pattern of the required type at the expense of as few failures as possible.

▸ *The pole-balancing system.* The first attempts on the problem were made with the help of Dean Wooldridge junior, who did all of the programming and contributed to the discussions of overall design. The physical apparatus chosen for our experimental trials is similar to one originally used by Donaldson [2] for a different, but related, purpose. It consists of a motor-driven cart free to run on a track of fixed length, with a pole balanced on it, as shown in figure 1. Control is *via* a switch with only two alternative settings, 'left' and 'right', so that the motor is always exerting its full force either in the one direction or in the other. The apparatus emits a succession of signals describing its state at regular intervals of time. Each signal consists of a vector of four elements x, $\dot{x}$, θ, and $\dot{\theta}$, corresponding to the position of the cart on the track, the velocity of the cart, the angle of the pole and the rate of change of the angle respectively. In place of a state vector, a 'failure' signal may appear, which terminates the sequence. This indicates that the apparatus has either run off the track or allowed the pole to fall, or both, so that the system must be set up afresh and a new trial started. The controller's task is to construct a mapping from the state vectors (without any knowledge of their physical interpretation) onto the two-valued control variable, in such a way as to eliminate the occurrence of failure signals. It should be stated at once that in all our experiments the cart and pole apparatus has been *simulated* by a separate part of the computer program. To build it in hardware would, we judged, have consumed time and effort without at this stage offering anything in return.

Let us first give the problem a conventional representation by plotting the four state-variables along 4 mutually orthogonal axes. A state of the apparatus will then correspond to a point in the 4-dimensional co-ordinate space so defined. Trial-and-error learning could in principle proceed by the program asking itself at each interval of time 'Have I been here before? If so, what did I do, and what was the consequence'? But with state variables measured on continuous scales there is an infinite number of points in state-space, so that the answer to the above question in practice will always be 'No, I have not been here before'.

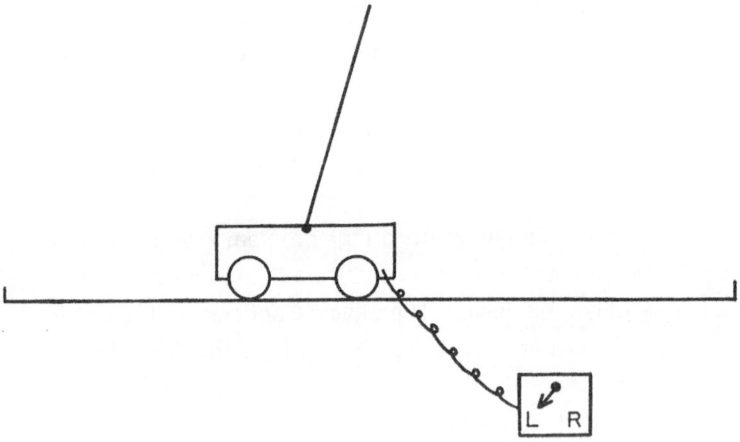

FIGURE 1 A representation of the apparatus to be controlled

We therefore took a leaf from the book of Widrow [3] who had been studying the cart and pole system from the same viewpoint as Donaldson (i.e. automatic learning by imitating a pre-existing controller, rather than *de novo* by trial and error). Widrow had divided the infinitely many possible states of the system into a relatively small number of classes by setting thresholds on the state variables, thus partitioning the total state space into a small number of sub-spaces, or 'boxes' as we shall call them. The partitioning which we ourselves adopted is shown below. Only three grades of position x, are distinguished, together with six grades of angle θ, three of velocity $\dot{x}$, and three of angle-change $\dot{\theta}$, giving 162 boxes altogether. The thresholds used are:

$x: \pm 7, \pm 21$ ins.,
$\theta: 0, \pm 1, \pm 6, \pm 12$ degrees,
$\dot{x}: \pm 1, \pm 30$ ins./sec.,
$\dot{\theta}: \pm 6, \pm 24$ deg./sec.,

and in each case, any excursion beyond the outer limits results in the immediate production of a failure signal.

▶ *A collective of automata.* We shall now describe our solution to the problem. As we do so a certain relevance will, we think, become apparent to the notion, previously aired, of delegating responsibility to a collective of lower-level automata each independently following a local decision rule.

In each box (*see* figure 2) we imagine an automaton, depicted in the diagram as a humanoid slave. This robot has a switch which he can set to signal 'left' or 'right'. He also has a scoreboard, divided into two columns, headed 'left' and 'right' respectively. In the left-hand column the slave keeps an up-dated estimate of the expected further life-time of the system until failure following a 'left' decision, and he obtains this estimate as a weighted average over all past occasions on which the system-state entered his box and he gave a 'left' decision. The right-hand column is entered correspondingly. So what happens when the

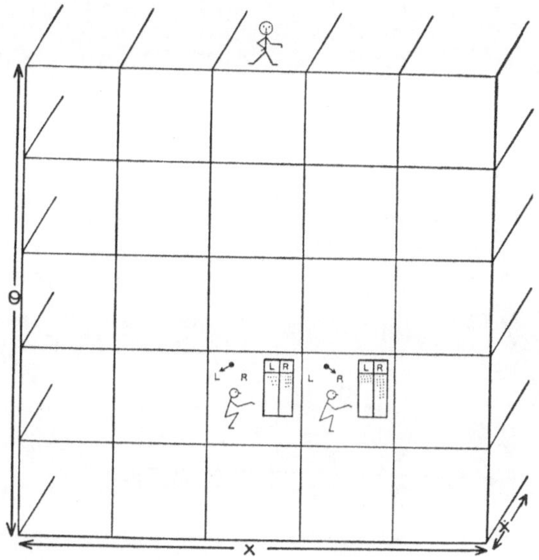

FIGURE 2
The state space (for clarity omitting the fourth, i.e. $\dot{\theta}$, dimension) divided into 'boxes' with an independent automaton in each box and a 'leader' supervising

computer program which implements this system is actually running? There is one 'top automaton' or 'leader', as shown in the figure, who inspects successive values of the state signal and hence always knows in whose box the state of the system is currently sited. If the state signal says 'failure' he relays the message to all boxes. Otherwise he alerts the slave whose box currently contains the system-state, and this slave proceeds as follows:

1. He compares his current values for 'expected lifetime following "left" decision' and 'expected lifetime following "right" decision' According to which is the greater, he sets his switch to 'left' or to 'right'.
2. He begins to count.
3. When he eventually receives a failure message from the leader, he uses his count to up-date either the left, or the right, column of his score-board.

This completes the outline logic of our adaptive control system. There are various further details, which we shall not discuss here since they are available elsewhere [4]. But in essence that is all. Two points should be made at this stage:

1. The program works. Figure 3 shows a successful learning run.
2. The task is non-trivial as a problem in automatic control. Conventional methods all depend upon fairly exact specification of the physical parameters of the apparatus, which in our case is withheld as part of the conditions of the

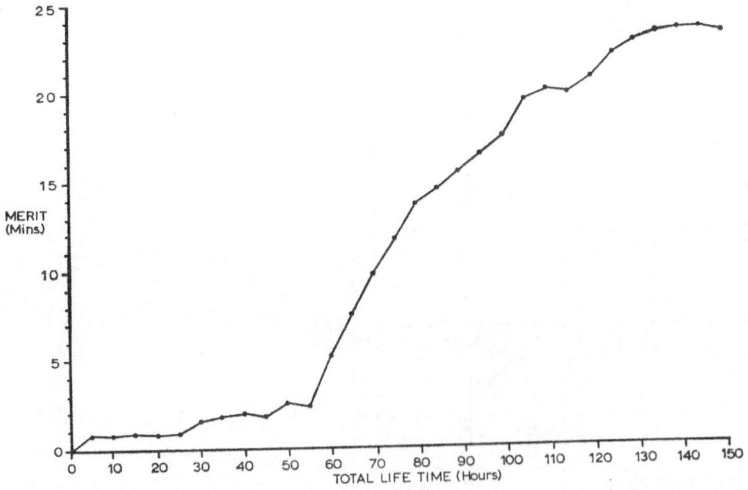

FIGURE 3
A successful learning run. The ability of the system, MERIT, is plotted against the TOTAL LIFE TIME over which the system has been learning. MERIT is a weighted average of the duration, from set-up to failure, of the control runs so far performed, giving most weight to recent runs

problem. The program is in fact written so that a simulation of *any* physical system with not more than 4 state variables can be plugged into it, there being no dependence whatever upon having a physical interpretation of the variables.

▶ *Pattern-formation.* Now for the biological analogies. These are rather wide-ranging. The principle of building up the pattern of a total system by exploiting the collective behaviour of locally autonomous sub-systems seems to be a characteristic strategy in Nature. Longuet-Higgins [5] has referred to 'sub-routines' in the context of morphogenesis, and we shall undoubtedly find this design feature in every complex system which we examine in detail, whether neuro-physiological,

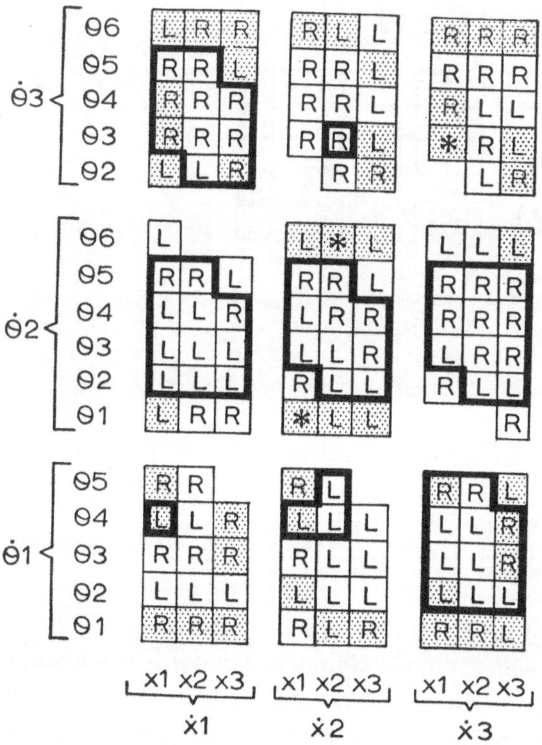

FIGURE 4

Final state of decision array with 'informed' boxes included within heavy line

KEY: R – 'right', L – 'left', * – extremely little information for decision

Plain boxes – symmetrical, Shaded boxes – unsymmetrical

$x1 - (-21 : -7), x2 - (-7 : +7), x3 - (+7 : +21)$ ins.

$\dot{x}1 - (-30 : -1), \dot{x}2 - (-1 : +1), \dot{x}3 - (+1 : +30)$ ins./sec.

$\theta1 - (-12 : -6), \theta2 - (-6 : -1), \theta3 - (-1 : 0)$ deg.

$\theta4 - (0 : +1), \quad \theta5 - (+1 : +6), \theta6 - (+6 : +12)$ deg.

$\dot{\theta}1 - (-24 : -6), \dot{\theta}2 - (-6 : +6), \dot{\theta}3 - (+6 : +24)$ deg./sec.

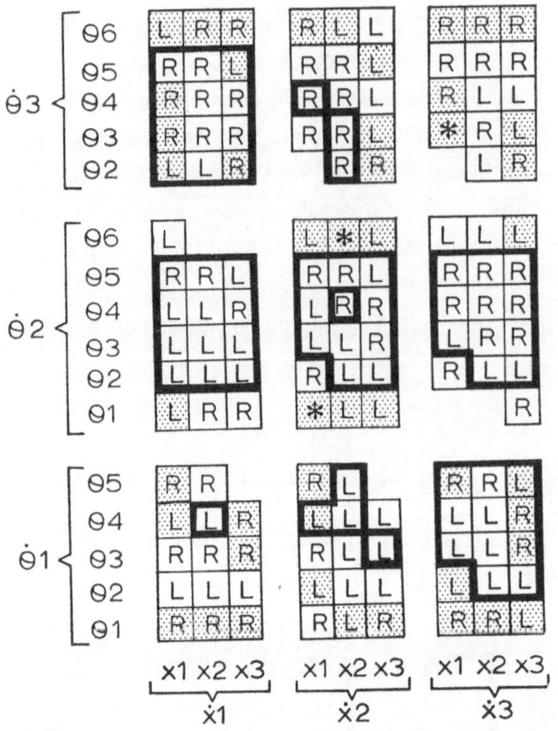

FIGURE 5
Final state of decision array with 'decisive' boxes included within heavy line
KEY: *see* figure 4

morphogenetic or evolutionary. But first a word to the effect that boxes and poles really have something to do with pattern-formation.

Consider the initial state of the 4-dimensional array of 162 boxes. The settings of the control switches contained in these boxes start in a random configuration: some point left, some point right – they make no pattern. However, the *final* state of the array, although still showing some randomness in areas where boxes have been rarely visited, presents other areas of quite elaborate pattern. Aspects of this pattern formation are depicted in figures 4 and 5 which display the array of final decision settings for a successful run, omitting the 'inaccessible' boxes which involve impossible combinations of the state variables. The version of the program used has monitoring facilities which enable inspection of the contents of the boxes at any stage of the run. These facilities were used at the end of the run and in the figures account is taken of the amount of information accumulated in each box and the relative value assigned to 'left' and 'right' decisions in each box. By taking a simple measure of each of these two factors, and for each box comparing the measurements obtained against chosen threshold values, the boxes are classed as 'informed' or 'uninformed', in figure 4, and as 'decisive' or 'indecisive', in figure 5. These classifications have a bearing on the two main pattern features shown in the figures, namely the occurrence of symmetry and the existence of 'switching boundaries'.

Since the task is symmetrical one might reasonably expect a system self-evolved to perform this task to show an analogous symmetry. The nature of this symmetry is as follows: by negating the co-ordinates of some point in the state space we produce a symmetrical point and if, say, a 'left' decision is necessary to maintain control at the first point then a 'right' decision will be required at the symmetrical point. Since a box represents a collection of similar points, this symmetry should carry over from points to boxes. We found, as expected, that the pattern formed in the array of boxes shows about 50% symmetry in the initial random configuration and an increasing degree of symmetry as boxes gain more information on which to base their decisions. Finally, in figure 4, we find that the 'informed' boxes show an 84% symmetry while the remaining 'uninformed' boxes still show only a 50% symmetry.

Compare now figure 4 with figure 5 and note the groups of boxes which are both 'informed' and 'decisive'. In these groups there are clearly demarcated regions in which all the switches are set to 'left', and other regions in which they are set to 'right'. These correspond to regions of the state space in which a 'left' (or 'right') decision is essential if control is to be retained. There are also 'don't care' regions in which the situation can still be retrieved regardless of whether a 'left' or a 'right' decision is given, although the case of retrieval may vary from place to place within such a region. Here, where boxes are 'informed' but 'indecisive' e.g. box ($x2$, $\dot{x}2$, $\theta4$, $\dot{\theta}2$), there will be some variability in the details of the patterns attained in different learning runs – rather as palm-prints vary from one individual to another in some regions of the hand, but show uniformity in others. Within the groups of 'informed' boxes these regions of 'left', 'don't care' and 'right' are clearly distinguishable and have quite simple switching boundaries.

The regions of 'uninformed' boxes are mainly those in which control is very difficult, if not impossible. For example, in both regions ($\dot{x}1$, $\dot{\theta}1$,) and ($\dot{x}3$, $\dot{\theta}3$) the velocity of the cart, $\dot{x}$, and the angular velocity of the pole, θ, have the same sign. To reduce $\dot{\theta}$ the cart must accelerate and so bring $\dot{x}$ nearer the failure level, while a deceleration to reduce $\dot{x}$ will send $\dot{\theta}$ nearer to the failure level. Each control run is started with the system at a randomly chosen point within the two centre boxes

$(x2, \dot{x}2, \theta 3, \dot{\theta}2)$ and $(x2, \dot{x}2, \theta 4, \dot{\theta}2)$, and can therefore only reach these difficult regions as the result of the decision of a more central box. Since 'uninformed' boxes are those which have only been visited infrequently, it can be seen that the system is evolving its own classification of 'controllable' and 'uncontrollable' regions, and learning to keep the system in the former region.

▶ *Splitting and lumping.* The weakness of our program in its present form is that it does not really live up to its ideal of independence of special knowledge of the physical apparatus. Some knowledge of this kind is in fact built into it when the user specifies the thresholds to be set on the state variables. It is easy to choose these in such a way as to make the control task impossible. In our plans for the next stage we aim to endow the program with the power to change the boundaries of boxes, by processes of 'splitting' and 'lumping'. Since there is a tempting analogy here with biological differentiation and de-differentiation we shall briefly sketch our approach, even though the new feature is not yet implemented.

After the boxes system has been allowed to run on its self-organizing mission for some time, it becomes reasonable to make a crude classification of boxes into 3 types as indicated earlier. Of these, two give no trouble. They are

1. boxes which have seldom been visited and hence have accumulated little evidence to discriminate between 'left' and 'right' settings and
2. boxes which have accumulated much evidence pointing clearly to either one or other decision.

The troublesome category contains

3. boxes with heavy accumulation of evidence which yet is ambiguously divided between the 'left' and 'right' alternatives.

Such an occurrence indicates that box boundaries have been set badly, for it hints that there are some points within one and the same box for which a 'left' is mandatory and other points at which a 'right' is demanded. Accordingly this is used as a signal to trigger the subdivision of the box concerned into 2 smaller boxes. The result will be that *in the vicinity of switching boundaries* state space becomes subdivided on a finer scale. Contrariwise, when two or more adjacent boxes are found to agree not only internally but also with each other, they are lumped into a single box. Consequently as the system evolves fine distinctions multiply and become concentrated where they are needed, and obliterated where they are not. How exactly the continuous switching boundaries will in practice be tracked by this process of discrete approximation remains to be seen.

In conclusion the adaptive system described possesses features in common with various self-organizing biological systems, notably evolution, morphogenesis and learning. In picturing the evolutionary parallel we must keep in mind that each box evolves in an environment which largely consists of the consequences of the other boxes' activities. The closest parallel would perhaps be with the small semi-isolated sup-populations of a large population undergoing unidirectional change. The degree to which these parallels clarify rather than confuse will, of course, vary from one person to the next. It is in the hope that some of our readers will be enabled to see old things in a slightly new way that we have ventured to report this work.

ACKNOWLEDGMENTS

This paper describes developments of work reported by one of us (D. Michie) at the Symposium on Biological Theory and Philosophy, Bellagio, in September 1966. Our thanks are due to Professor C. H. Waddington F.R.S. who organized the Symposium and to the Rockefeller Foundation who generously acted as hosts.

References

1. Moore, E. F. (1964). The firing squad synchronization problem. *Sequential machines: Selected papers.* (ed. E. F. Moore). Addison-Wesley Publishing Co. Inc.: Massachusetts. 213–14.

2. Donaldson, P. E. K. (1960). Error de-correlation: a technique for matching a class of functions. *Proc III. International Conf. on Medical Electronics.* 173–8.

3. Widrow, B. and Smith, F. W. (1964). Pattern recognising control systems. *Computer and information sciences.* (eds. J. T. Tou and R. H. Wilcox). Clever Hume Press. 288–317.

4. Michie, D. and Chambers, R. A. (1968). BOXES: an experiment in adaptive control. *Machine Intelligence 2* (eds. E. Dale and D. Michie). Oliver and Boyd: Edinburgh. 137–52.

5. Longuet-Higgins, H. C. (1966). Personal communication at the Bellagio Symposium.

Note on Topics for the Second Symposium

Brian Goodwin
University of Sussex

Physics deals primarily with systems describable by time-invariant quantities, among which energy is central. Biology deals with systems in which process and change are the outstanding characteristics. This is not to imply that there are no time-invariant quantities which could define the 'envelope' within which the system moves, but it is clear that energy is not one. Invariants for systems that do not conserve energy obviously exist, as for example Volterra's invariant for prey-predator systems. They conserve something else. It could be argued (albeit largely on intuitive grounds) that the property of life should be definable as a constant characteristic of a biological system; that some function of motion, activity, and change should represent biological process, as energy represents physical process. However, it is certainly not possible to force biological behaviour into laws of quantitative invariant form if it doesn't want to go. It would be prejudging the issue to assume that the task of theoretical biology was to discover and define such laws. This might be not only ultra-ambitious but also misguided.

A more cautious and empirical approach would be to focus attention on certain key concepts which kept recurring at last year's symposium. Possibly a fairly exact definition of these concepts and how they relate to one another would be a useful partial goal for this year's meeting. Some of these concepts with questions on their relationships are set out below in what appeared to be a natural cycle, starting and ending with energy.

(1) *Energy and order*

1. What kind of order can be deduced from considerations of energy interactions between units?

2. What types of biological systems display such order? Virus coats, collagen matrix, embryonic cell layers?

3. What are the stability characteristics of such systems, i.e. what is the stability criterion?

(2) *Order and information*

1. Is it biologically useful to relate order (or disorder) and information (negentropy) in the usual manner,

$$I = -k \log \Omega = -S,$$

where, for example, Ω = number of complexions of a macromolecule, or number of positions an embryonic cell can occupy?

2. Does this definition of information express what we mean when we talk of information in the context of control, e.g. the control of enzyme activity by small molecules?

3. What is the relation between information and specificity?

(3) *Information and organization*

1. Does organization necessarily involve information exchange between parts?

2. Can organization be defined in terms of informational interaction between units, in analogy with definitions of order in terms of energy interactions?

3. What stability criterion defines organized behaviour? Adaptation, survival? How does this differ from physical stability?

(4) *Organization and memory*

1. Any organized system is adaptive in relation to some environment. Does this adaptation necessarily involve prediction, hence memory? Thus is memory a necessary feature of an organized system?

(5) *Memory and self-reproduction*

1. Is memory necessary for self-reproduction?

2. What is the simplest automaton capable of self-reproduction? A finite state automaton? A Turing machine?

(6) *Self-reproduction and measurement*

1. Are quantum mechanical questions of measurement relevant to the problems of self-reproduction in macromolecules like DNA?

2. Are there any difficulties encountered in reconciling quantum mechanical principles with the apparent reliability (stability) of macromolecular synthesis?

(7) *Measurement and energy*

1. Can biologists be satisfied with the physical dogma that all systems must be ultimately describable (in principle) by quantum mechanics, hence in terms of physical observables such as energy, until such physical dilemmas as those of quantum mechanical measurement are resolved?

2. Would the resolution of this problem have any relevance to biological theory?

Comment by C. H. Waddington

The gist of the points I want to make are in Brian's first paragraph, and I do not believe he is right in suggesting that we should weaken and settle for trying to formulate theories in the terms of conventional physics.

It seemed to me that what emerged from the first meeting was that biology has three major characteristics:

(i) It *essentially* involves time at three scales: turnover, development of a phenotype, evolution; there can be no instantaneous life, nor life that merely undergoes cyclic or reiterative development.

(ii) It is *essentially* multi-dimensional: there can be no unit, however complex its internal structure, which can be considered 'living' without reference to the situation not included in the unit.

(iii) It is *essentially* organized, in the sense that it exhibits homeorhesis (not homeostasis): there can be no living thing which can vary equally easily in all conceivable directions as time passes.

It would follow from (i) that the basic concepts of biology cannot be dimensionless, like energy or information, but must be such that time cannot be omitted from them. In place of energy, or even velocity (which is dimensionless if space and time are equivalent), we must have some analogue of acceleration. In place of information, algorithm.

From (ii) it follows that the basic theory must have the form of a topology of an n-dimensional space.

From (iii) it follows that this topology is chreodic – a progression along stabilized time-trajectories.

It does not seem to me *a priori* impossible to develop a logic (or mathematics) whose basic concepts are of the type required by (i) and (ii) and which issues in functions having the property (iii). And in my view, nothing less than this is worth calling 'theoretical biology'.

To some extent, I think, this point of view is incorporated in Brian's question, whether organization necessarily involves memory (though I should like to include anticipation as well as memory, i.e. duration).

Comment by H. H. Pattee

Dear Professor Waddington,

Here are a few thoughts on your comment on Goodwin's note outlining the second Bellagio programme.

I agree with you that we should not 'weaken' and use a 'cautious' approach. The problem of life certainly requires a daring full-scale attack on all fronts. Your three points which emerged from the first meeting are consistent with my own evaluation. However, I would like to elaborate on them, first to see if I understand what you mean, and second to connect them up with what I mean.

Point (i). *Multiple time scales are certainly a crucial aspect of life. Physics commonly uses only one time scale (except in some non-linear thermodynamics) I would only say that three scales may not be enough to describe the whole of biology. For example, there is physical time as in the equations of motion, there is catalytic time which may be necessary to describe enzymes, there is cellular reproduction time, there is organism development time, there is individual generation time, there is ecological succession time, and finally there is evolutionary time. Perhaps psychological or conscious time should also be added.*

Point (ii). *There is no living unit which can be considered 'living' without reference to the external environment. Here I would say that your use of 'multi-dimensionality' and 'topological' are not as descriptive as biological terms. Biologists should emphasize over and over that 'living' is unavoidably a* total ecosystem *property and not the property of an isolated collection of macro-molecules. It seems to me that the central question of the origin of life is* not, *'Which comes first, DNA or proteins?', but rather 'What is the simplest possible ecosystem?' I think Cairns-Smith would agree.*

Point (iii). *You say that life's 'topology is chreodic', and that you think it is possible to develop a logic whose basic assumptions are consistent with points (i) and (ii) but whose consequences have the chreodic property. I think I agree with your intuitive picture of the situation, but let me try to clarify the difficulties that I see.*

Words like homeorhetic and chreodic suggest to me the deterministic response of a very complicated system to a very complicated environment, which never-theless in its logical elements is basically an hereditary *response. That is, a choice or selection or decision is made between physically possible alternatives. (If there is more than one alternative and you choose to follow only one, then you have performed a many-one transformation.) Physically, a device which makes such a transformation, like a gate or switch, is necessarily a non-holonomic constraint, which simply means that there are more dimensions in the space which can describe possible systems than there are dimensions to the actual trajectory or development of a given system. If you wish, you could also call this an 'organized' system, or a system with a memory.*

Now the problem is that elementary physical laws use only one time scale and express only one-to-one or 'group' transformations on the variables. (Incidentally, I think that 'topological' mappings are also only one-to-one. 'Semi-group' or automata transformation are many-one.) What do we do with this problem? We can certainly invent a 'mathematical biology' which formally describes or predicts biological behaviour'. But a 'theoretical biology' means more than that to me. It must be a 'physical biology' which does not ignore the physical basis of all our observations and interactions with the world. Let me emphasize that I consider a physical biology as a likely possibility but by no means easy to develop.

Goodwin's note asks, 'Can biologists be satisfied with the physical dogma that all systems must be ultimately describable by quantum mechanics . . .?' That is not a fair way to ask the question. It is mostly the new biologists like Crick, Watson, and Kendrew who say they are satisfied with this dogma; it is the physicists like Bohr, Schrödinger, and Wigner who know better. I believe that both biologists and physicists cannot be satisfied with this dogma, but more to the point, they may not ignore the tension between physical and biological descriptions. If biologists ignore physics they only work with mathematical models which may give the right answers but for the wrong reasons.

I think, then, that I agree with your picture of biology. But to 'weaken' our efforts in theoretical biology is to forget about physical laws and settle for only a 'mathematical' biology. In fact, I shall support the view that the secret of life is hidden in this tension between physical and biological description. This is at the root of the traditional problems of object and subject, determinism and freedom, matter and mind, dead and living. Of course it is presumptuous to hope to solve such problems once and for all, but it is also futile to ignore them in the context of a theoretical biology.

I shall save further comments for the Villa.

Sincerely yours,

Howard

Letter from Elsasser to Pattee

Dear Howard,

After returning here I found a letter which you had addressed to Professor Waddington dated July 6 and which Waddington apparently then distributed to members of the conference.

As I hardly have to tell you, I am on the whole very much in agreement with you, but there are a few minor points on which I might as well comment now. They might not have been taken up at the conference in any event.

The first is that I firmly object to putting Bohr, Schrödinger, and Wigner into one pigeon-hole – Bohr says that the characteristic feature of the autonomy of any natural phenomenon is that when one approaches the system from two quite different directions one is likely to obtain mutually contradictory answers. He also says that he thinks quantum mechanics is not yet adequate enough to describe life, but only in the sense of a more extended complementarity. What I myself have tried to show is that such a broadening is not likely to involve a change in the mathematics of quantum theory. I have known Schrödinger quite well and know Wigner, and I believe it is only fair to say that neither of them has given even remotely the extent and depth of thought to relationships of physics and biology as Bohr has. His father was not only a biologist but, so I have recently been assured, a very excellent and thoughtful one at that.

I could not agree with you more about the greater depth that would be involved in 'theoretical' biology than in 'mathematical' one. On the other hand, as you yourself remarked, one then gets oneself much deeper into 'philosophy'. I think this is quite simply because we are all organisms and so biology is really much closer to our hearts than physics is. However, I think that the concepts mentioned by you, namely object and subject, determinism and freedom, matter and mind, are essentially metaphysical in this form. This would have been all right before Whitehead and Russell came along. They taught us that one must formulate science in terms of purely abstract structures which then ought to be matched to data of experience.

I think the main problem in this connection when it comes to biology is the perfectly valid objection made by many people against Darwinism, that ultimately evolution must include some form of creativity. But Henri Bergson made a colossal blunder when he tried to introduce the concept of creativity, which ought to remain metaphysical, directly into his discourse. I have tried to avoid this carefully by reformulating biology on introducing 'unanswerable questions' instead. This is a little bit after the manner of the philosopher Wittgenstein, who once said: 'Where there is no answer, there is no question.'

My personal feeling from the rather close contact with biologists which I have had in recent years is that we might well see one or two generations preoccupied with mathematical biology before one sees a real theoretical biology arise. So we have to be patient. Certainly the Bellagio conference has greatly enhanced the clarity of ideas (at least of my ideas). I wish I had had this opportunity before and not after completing my book. On the other hand, the conference contained a large fraction of those interested in genuine theoretical biology, so we will have to go slow in any event.

Please note the enclosed change of address. With best personal regards.

Sincerely yours,

Walter M. Elsasser

List of Participants

with references to some works of particular relevance to the Symposium

A. G. **Cairns Smith**	Chemist, specializing on 'artificial enzymes' and origin of life problems. See *The Origin of Life and the Nature of the Primitive Gene* (J. Theoret. Biol., 1966, *10:* 53). Department of Chemistry, University of Glasgow, Scotland.
Sam **Devons**	Physicist. Editor, *Columbia Symposium on the Relations of Physics to Biology* (Columbia University Press, ready shortly). Department of Physics, Columbia University, New York, U.S.A.
Jack **Cowan**	Mathematical Neurobiology. Formerly Imperial College, London; from October, 1967, Head of Department and Professor of Mathematical Biology, University of Chicago, U.S.A.
Brian **Goodwin**	Mathematical Biology, particularly limit cycles in metabolic networks, and biological periodicities. See *Temporal Organisation of Cells* (Academic Press, 1964). Department of Biological Sciences, University of Sussex, Brighton, Sussex.
Karl **Kornacker**	Biomathematics. 56–526, M.I.T., Cambridge, Mass., U.S.A.
H. **Kroeger**	Experimental Biologist, chromosome puffing, and epigenetics of pattern. Eidgennossische Technische Hochschüle, Zürich, Switzerland.
Richard C. **Lewontin**	Evolutionary Biology. Department of Zoology, University of Chicago, U.S.A.
Paul **Lieber**	Theory of 'classical' physics. Department of Mechanics, University of California at Berkeley, California, U.S.A.
C. **Longuet-Higgins**	Formerly Professor of Theoretical Chemistry, University of Cambridge, England; since 1967 Professor of Machine Intelligence, University of Edinburgh, Scotland. Fellow of the Royal Society.
J. **Maynard Smith**	Evolutionary genetics, Senescence, Pattern Formation. Head of Department of Biological Sciences, University of Sussex, Brighton, Sussex, England.

Ernst **Mayr** Evolutionary Biology. His many theoretical works are summarized in a major and classical book *Animal Species and Evolution* (Harvard University Press, 1963). Member U.S. National Academy of Science, Agassiz Professor, Museum of Comparative Zoology, Harvard University, Cambridge, Mass., U.S.A.

Donald **Michie** Formerly experimental biologist (mammalian reproductive physiology, immunology), now Professor of Machine Intelligence Programming, Edinburgh University. Department of Machine Intelligence and Perception, University of Edinburgh, Scotland.

Howard H. **Pattee** Physicist, particularly concerned with polymerization processes. See *Physical Theories, Automata, and the Origin of Life*, in *Natural Automata and Useful Simulations*, ed. Pattee and others (Spartan Books, Washington, D.C. 1966). Department of Biophysics, Stanford University, California, U.S.A.

John R. **Platt** Biophysicist. See *The Step to Man* (Wiley, New York, 1966) and *Properties of Large Molecules that go beyond the Properties of their Chemical Groups* (J. Theoret. Biol. 1961 vol. 1, 342). Now Director of Mental Health Research Institute, University of Michigan, Ann Arbor, Mich., U.S.A.

Ruth **Sager** Molecular genetics, specializing in non-chromosomal heredity. Co-author of the first classical textbook of molecular biology, Sager and Ryan, *Cell Heredity* (Wiley, New York, 1961). Department of Genetics, Hunter College, New York, U.S.A.

René **Thom** Topologist. His book *Stabilité Structurelle et Morphogénèse* is shortly to be published by Benjamin, New York. Awarded Field Prize for Mathematics. Institute des Hautes Etudes, Bures sur Yvette, Seine et Oise, France.

C.H. **Waddington** Embryology, Genetics, Evolution. Has published many books on theoretical biology, from *Organisers and Genes* (Cambridge University Press, 1940) onwards; particularly, *The Strategy of the Genes* (Allen & Unwin, London, 1957), *The Ethical Animal* (Allen & Unwin, 1960, Chicago University Press paperback 1967), *New Patterns in Genetics and Development* (Columbia University Press, 1962), and contributions to *Mathematical Challenges to the Neo-Darwinian Interpretation of Evolution* (Wistar Institute Symposium Monograph, 1967). Fellow of the Royal Society. Institute of Animal Genetics, Edinburgh, Scotland.

Lewis **Wolpert** Developmental biologist, trained as engineer. Department of Biology, Middlesex Hospital Medical School, London, England.

Christopher **Zeeman** Topologist. See Topology of the Brain in *Mathematics and the Computer Services in Biology,* Medical Research Council, 1965. Department of Mathematics, University of Warwick, Coventry, England.

Author Index

References to extended treatments are given in italics

Author Index

Author Index

Subject Index

Subject Index

Subject Index

Subject Index

Subject Index